D0835076

Get a Grip on
PHYSICS

John Gribbin

DOVER PUBLICATIONS, INC.
Mineola, New York

Copyright

Bibliographical Note

This Dover edition, first published in 2011, is an unabridged republication of *Get a Grip on New Physics,* originally published in 1999 by Weidenfeld and Nicolson, London.

Library of Congress Cataloging-in-Publication Data

Gribbin, John R.
[Get a grip on new physics]
Get a grip on physics / John Gribbin. — Dover ed.
p. cm.
Originally published: Get a grip on new physics. London : Weidenfeld and Nicolson, 1999.
Includes bibliographical references and index.
ISBN-13: 978-0-486-48502-7 (alk. paper)
ISBN-10: 0-486-48502-1 (alk. paper)
1. Physics—Popular works. I. Title.

QC24.5.G75 2011
530—dc23

2011016863

Manufactured in the United States by Courier Corporation
48502102
www.doverpublications.com

CONTENTS

INTRODUCTION

THE OLD PHYSICS

there's a rule for everything, and everything has its rule

Isaac Newton

* Modern physics - or, at least, the first phase of modern physics - began with Isaac Newton, in the second half of the 17th century. The most important thing Newton did was to spell out that the entire Universe is governed by simple rules, which also apply to things going on here on Earth. The most famous example of this is his LAW OF GRAVITATION, which explains both the way an apple falls to the ground from a tree and how the Moon stays in orbit around the Earth - and much more besides.

OLD AND NEW PHYSICS

Old physics is the stuff we learn in school, the kind of laws that apply to objects we can see and touch, like billiard balls or cars. New physics deals with things that are inaccessible to our senses, like atoms and black holes.

Newton's telescope

NEWTON AND GRAVITY

* *This law of nature is what is known as an* INVERSE-SQUARE LAW *– the force of attraction between two objects depends on their two masses multiplied together, divided by the square of the distance between them.* So if the same two objects are twice as far apart the force is reduced to a quarter, while if they're three times as far apart it is reduced to a ninth. And so on.

* But, for the moment, the law itself is less important than the fact that there is a unique law that describes the force of GRAVITATIONAL ATTRACTION operating between any two objects in the Universe –

between a pencil on my desk and the cat in the next room, between the Moon and the Earth, or between two galaxies on opposite sides of the Universe, or even between my cat and a distant galaxy.

BEFORE NEWTON

* Before Newton came along, even scientifically minded people commonly believed the Universe was governed by rules devised by the gods, or God. When, in 1609, **Johannes Kepler** realized that something made the planets stay in orbit around the Sun, he called it the 'Holy Spirit Force', and nobody laughed at him for doing so. ***The Universe was seemingly at the mercy of mysterious and incomprehensible forces, which might change from day to day or from place to place.***

KEY WORDS

GRAVITY:
the force of attraction between two masses

GRAVITATION:
the influence any object exerts on other objects in the Universe simply by having mass

Johannes Kepler (1571–1630)

German astronomer who discovered the laws of planetary motion, which helped Newton develop his theory of gravity. Kepler used observations of the planets compiled by **Tycho Brahe** (1546–1601). Before joining Brahe in Prague, he trained for a career in the Church, then worked as a teacher of mathematics at a Protestant seminary in Graz.

Isaac Newton (1642–1727)

Newton was active in many fields. He studied alchemy (still almost respectable at that time) and theology, and served as a Member of Parliament (his knighthood was for political work, not science) and as Master of the Royal Mint and President of the Royal Society. Newton was very secretive about his work, and often got involved in huge rows with other scientists about who had thought of an idea first (usually he had thought of it, but hadn't bothered to tell anyone!). His great work in physics was completed before he was 30, but only published in 1687, at the urging of Edmond Halley. In the 1690s Newton suffered a mental breakdown, and although he recovered sufficiently to lead a normal life, he did no more scientific work.

Edmond Halley

A CLOCKWORK UNIVERSE

tick tick tick

* After Newton, the Universe was perceived in a quite different way - as a kind of cosmic clockwork mechanism, running predictably in accordance with laws of physics that could be determined from experiments here on Earth. The laws might be God-given (Newton thought they were), but they were now seen as being the same everywhere and at all times.

Newton identified the laws of the Universe

Newton suffered a breakdown from overwork

FUNDAMENTAL LAWS

* ***The predictability of the Newtonian Universe was based on three other fundamental laws discovered by Newton.*** Known as Newton's LAWS OF MECHANICS (or laws of motion), they are spelled out in his great book *Philosophiae Naturalis Principia Mathematica (The Mathematical Principles of Natural Philosophy)*, usually referred to simply as the Principia.
* ***These three laws formed the basis of physics for the next 200 years – and still suffice to explain the way things behave in the everyday world, even though some of the things explained by them (such as the flight of a jet aircraft, or the journey of a space probe to the planet Jupiter) were undreamed of by Newton himself.***

Newton's first law of mechanics

The first of Newton's three laws of mechanics immediately shows how physicists often have to discount 'common sense' in order to get a grip on the way the world works. ***It insists that any object – by implication, any object in the entire Universe – either stays still or keeps moving in a straight line unless some force is applied to the object.*** The standing still part is no problem, so far as common sense is concerned. Here on the surface of the Earth most things do stay still, unless they are given a push. But if given a push, they certainly don't keep moving in a straight line for ever. They slow down and come to a halt.

KEY WORDS

MECHANICS: the branch of physics that deals with the way things move and the forces that make them do so

ON AND ON AND ON...

* The first step in Newton's insight was to realize that things only come to a halt because they are being influenced by an outside force - the force of friction. Things stop moving because they are rubbing against other things, even if the other things are only molecules of air brushing past.

Galileo realized the balls would keep on rolling unless something stopped them

UNSTOPPABLE

Imagine something sitting in empty space then given a quick push (a fairly obvious thing to imagine today, in the age of space flight, but a huge leap of the imagination in the 17th century). It will keep moving in a straight line for ever unless some other force acts on it.

...FOR EVER

* This business about friction bringing things to a halt was, in fact, already partly understood before Newton came onto the scene. In particular, **Galileo Galilei** had realized that things would keep moving for ever if no external force acted on them. He came to this conclusion after carrying out a series of experiments in which balls were rolled down inclined planes. *The balls rolled off towards the horizon, and Galileo realized that without friction they would keep rolling for ever.*

who knows where this ball will end up

...OR ROUND AND ROUND

★ ***At this point Galileo made a daring but erroneous extrapolation.*** Like all educated people of his day, he knew that the Earth is round. So an object that keeps moving towards the horizon for ever must be following a circular path around the surface of the Earth, and will eventually end up back where it started from. A least it would do if there were no mountains or other obstructions in its way. ***This led Galileo to believe there was a fundamental law of nature which said that, left to their own devices, things move in circles.***

Galileo was tried for proclaiming that the Earth went round the Sun

Galileo Galilei

Galileo Galilei (1564–1642)

Galileo was the first person to use a telescope to observe the stars and planets scientifically. He studied medicine at the University of Pisa, but dropped out to become a scientist. His astronomical observations made him famous, and he was one of the first scientists to publicly support the idea that the Earth goes round the Sun. As a result, when 69 years old and in frail health, he was tried for heresy, forced to recant under threat of torture, and confined to house arrest for the rest of his life. The publicity of his trial in Catholic Rome helped to ensure that his ideas were taken up in Protestant northern Europe.

WOW!
it's going round in circles

CIRCULAR THINKING

* Don't believe everything you see on TV. Those spaceships that appear to be steering a straight course are actually in orbit around the Earth, moving along more or less circular paths. And the things moving 'in straight lines' inside the spaceships - that is, in straight lines relative to the walls of the spaceship - are also circling the Earth.

orbiting spaceships follow a circular path

Heretical thinking

Nicolaus Copernicus (1473–1543) was a Polish astronomer who was the first scientist to promote the idea that the Earth goes round the Sun.

Nicolaus Copernicus

NEW, EXCITING AND HERETICAL

* Galileo would have quite happily accepted those TV pictures as evidence in favour of his argument. Indeed, the idea that circular motion was the natural order of things would have seemed particularly convincing to Galileo and the better educated of his contemporaries because of a relatively new, exciting and (literally) heretical idea, proposed by **Nicolaus Copernicus**, that the planets – including the Earth – move in circles around the Sun.

ALL BECAUSE OF GRAVITY

★ ***When Newton said that the natural order of things in the Universe is for objects to move in straight lines, he had to explain why the planets stay in orbit around the Sun and don't fly off into space.*** This is where his law of gravity came into the picture, not only explaining how the Sun maintains a grip on its family of planets, but also why the orbits of the planets around the Sun are – as Johannes Kepler had discovered, in 1609, to the embarrassment of Galileo – actually elliptical, not circular.

★ ***It is all thanks to Newton's inverse-square law of gravity. Stated more fully, this law says that the force of*** ATTRACTION ***operating between two masses is equal to the two masses multiplied together, all divided by the square of the distance between them (hence 'inverse square') and then multiplied by a constant, known as the constant of gravity, which is the same everywhere in the Universe and at all times.***

★ The only way to find out the constant of gravity, which tells you the strength of the force of gravity, is by experiments – but once you know this constant everything else is easy to calculate.

Saturn

KEY WORDS

ATTRACTION:
any force that pulls two objects together

MASS:
the amount of matter there is in an object

GRAVITY SIMPLIFIED

The simplest way to picture the effect of gravity is to imagine a stone, tied to a string, being whirled round and round in a circle. The analogy isn't exact, because the stone is moving in a circle, not an ellipse. But the force acting along the string is just like the force of gravity: it pulls the stone inwards and keeps it 'in orbit'. Should the string break, the stone would fly off in a straight line, at a tangent to its 'orbit'.

NEWTON'S SECOND LAW

*** Newton's second law of mechanics also comes into the picture. The second law tells you how much the motion of an object is affected by a force applied to it. It says that a force applied to a mass causes an acceleration.**

KEY WORDS

ACCELERATION:
any change in the speed of an object or the direction in which it is moving, or both
VELOCITY:
the speed of an object measured in a specific direction
FREE FALL:
state of weightlessness felt (or not felt!) by any object moving under the influence of gravity only

ACCELERATION AND VELOCITY

* ACCELERATION ***means a change in the*** VELOCITY ***of an object.*** And velocity – which is speed measured in a certain direction – has two properties. When the velocity changes, it may mean that the speed changes, as when you apply the brakes and bring a car to a halt in a straight line. Or it may mean that the direction of motion changes, as when you turn the wheel and take the car round a bend (or when a stone tied to a string whizzes round in a circle).

* ***So a change in velocity may involve a change in speed without any change in direction, or it may involve a change in direction without any change in speed, or it may involve a bit of both. They are all accelerations.***

FIRE!

if a cannonball is fired powerfully enough, it will travel right round the planet...

CURVING CANNONBALLS

* ***Newton himself made an analogy with a superpowerful cannon fired from the top of a tall mountain. Ignoring the effects of friction, imagine firing cannonballs off horizontally with increasingly powerful blasts from the cannon.***

* The first ball flies a little way towards the horizon and falls to the ground, tugged towards the centre of the Earth by gravity (it actually follows a curving, parabolic, path from the mouth of the cannon to the ground). The next ball travels a little further before gravity is able to pull it to the ground, and so on.

* ***But remember that the Earth is not flat – it curves away under the flying cannonball, which is, of course, always being accelerated towards the centre of the Earth. Because the surface of the Earth is curved, the cannonballs fly further over it than they would if the Earth was flat.*** (Indeed, firing cannonballs off like this and measuring how far they travel would be a way of proving that the Earth is not flat!)

Good shot!

If the cannon is capable of producing a powerful enough blast, the flying cannonball will travel right round the planet and hit the rear of the cannon. It will have gone into orbit. ***Because it moves forward and falls sideways all the time, the sideways fall is exactly enough to keep it in orbit – in a state that is sometimes described as 'free fall'.***

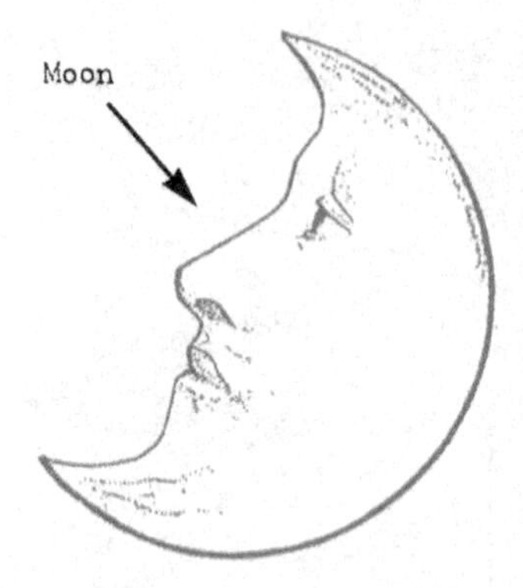

CANNONBALLS AND THE MOON

★ Like the cannonball, the Moon is always falling and always accelerating, even though its speed in its orbit does not vary significantly.

The tug of gravity

The orbit of the Moon around the Earth is a good example of acceleration at constant speed. The Moon would 'like' to keep moving in a straight line, but every time it moves forward, even a tiny bit, the force of the Earth's gravity tugs it sideways, deflecting it from a straight course. This constant sideways tugging keeps the Moon in its orbit.

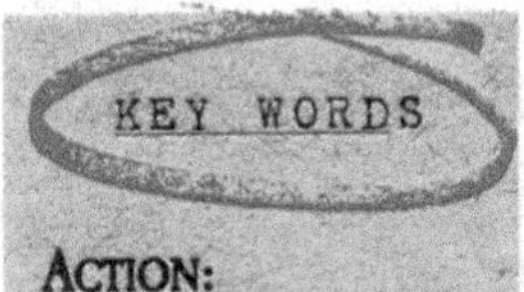

ACTION:
(in Newton's laws) a force – something that 'acts' on something
REACTION:
the force that pushes back when an action acts on something

THAT APPLE AND THE MOON

★ Newton's second law says that when a force *F* is applied to a mass *m* it causes an acceleration that can be expressed as $a = F/m$. ***The bigger the force applied to the mass, or the smaller the mass subjected to the same force, the bigger the acceleration produced.***

★ This second law of motion combined with the inverse-square law of gravity explains both the acceleration produced in an apple falling from a tree at the surface of the Earth and the acceleration of the Moon falling sideways in its orbit around the Earth. ***In both cases, the cause is the same – the Earth's gravity.***

NEWTON'S THIRD LAW

★ Newton's third (and last) law of motion can also be understood by thinking about what happens when a cannon is fired. The cannonball goes out of the mouth of the cannon and off into the distance, while the cannon itself rolls

the Earth's gravity keeps the Moon in orbit

backwards in the opposite direction. Similarly, when you fire a rifle you feel a kick as the rifle recoils.

★ ***Using the word 'action' where we would probably say 'force', Newton pointed out that for every*** ACTION ***there is an equal and opposite*** REACTION. A masterpiece of brevity, this law contains a large amount of information. First, it states that if you hit something, it hits back. This is quite easy to test. If you thump your fist on the table you can feel the reaction, quite unambiguously. The law also says that the action and the reaction are simultaneous.

Newton's third law of motion

EQUAL, OPPOSITE AND INSTANTANEOUS

Newton's third law also points out that the reaction is equal and opposite to the initial force. In spite of this, the cannon only recoils a little bit, while the cannonball goes off into the distance – because the same force is being applied to a more massive object, and the acceleration produced is, remember, inversely proportional to the mass it is applied to. But the forces themselves (the action and reaction) always cancel out precisely.

ACTION AND REACTION

★ There's more to action and reaction than you might think. The equality of action and reaction applies all the time: to yourself and everything around you. You are being pulled downwards by the Earth's gravity, which gives you your weight - but you are also pulling the Earth upwards by the same amount.

How a rocket works

Newton's law of action and reaction also explains how a rocket works. The rocket motor fires exhaust gases out in one direction, and this produces a reaction which pushes the rocket in the other direction. There is no need for the exhaust gases to have anything to push against – which is why rockets work in the vacuum of space. ***All that matters is that hot gases are squirted one way, and the rocket heads the other.***

THE EARTH MOVES?

★ If you were falling towards the ground from the top of a tall building, ***the Earth would also be moving up to meet you*** (but only by a tiny amount, since the acceleration depends inversely on the mass).

★ ***When you stand still on the ground, your weight is a force pressing downwards on the ground – which responds by pushing upwards with an equal and opposite force, keeping you in place.***

Albert Einstein

THE FUNDAMENTAL THINGS APPLY...

★ As the foregoing examples show, Newton's laws are still the fundamental principles on which physics operates for everyday purposes – on the scale of human beings, or even planets and stars.

★ But on a much larger scale – when we are talking about very massive objects, or the whole Universe – Newton's law of gravity is not quite good enough and we have to use the ideas developed by Albert Einstein in the 20th century, in his general theory of relativity. On a much smaller scale (smaller than atoms), Newton's laws of mechanics are not quite good enough either and we have to use another theory developed in the 20th century, quantum mechanics.

FROM STARS TO ATOMS

These two great ideas, relativity and quantum mechanics, form the basis of the new physics. But it is a sign of just how powerful Newtonian physics is that it applies very accurately to the behaviour of everything from stars to atoms, even though hardly anything was known about atoms when Newton was alive. ***Indeed, one of the greatest achievements of Newtonian mechanics is the way in which it has been used to explain the behaviour of gases, liquids and solids in terms of atoms and molecules, which move about and collide with one another in perfect obedience to Newton's laws.*** This second flowering of Newtonian theory happened in the second half of the 19th century – two centuries after he wrote the *Principia*.

CHAPTER 1

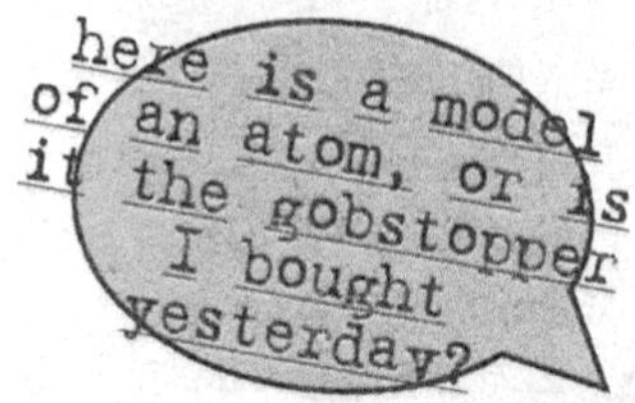

ATOMS AND MOLECULES

* An atom is the smallest unit of an element that can exist. The most appropriate image of it is a tiny hard sphere, like a minute billiard ball. Some substances in the everyday world (such as pure gold) are made of only one kind of atom. A pure-gold ring, for example, simply contains billions and billions of gold atoms.

H_2O

LINKING UP

* In some elements, identical ATOMS join together to form MOLECULES. This happens in the case of hydrogen, where each molecule is made up of two hydrogen atoms and is written as H_2. Other substances, such as water, are made of two or more different types of atom combined with one another to form molecules. The symbol for a hydrogen atom is H and the symbol for an oxygen atom is O – so, since two hydrogen atoms combine with one oxygen atom to form a molecule of water, the symbol for a molecule of water is H_2O.

* When they are on their own, oxygen atoms also like to link up with one another – so that the most common form of oxygen, including the

Tragic genius

Austrian physicist **Ludwig Boltzmann** (1844–1906) played a key role in developing the kinetic theory of gases, thereby helping to establish, albeit indirectly, that atoms are real. He became clinically depressed, partly because the atomic theory came under attack in his native Austria, and killed himself in 1906 – just a year after Einstein's work had, unknown to Boltzmann, proved the existence of atoms.

stuff we all breathe, is O_2. ***For the moment, though, all that matters is that these atoms and molecules can all be pictured as tiny balls, constantly in motion, bouncing off one another.***

the word kinetic comes from the Greek for motion

HOW GASES BEHAVE

* The people who worked out the details of this image of a gas as molecules in motion were **James Clerk Maxwell**, in Britain, and **Ludwig Boltzmann**, in Germany, in the mid-19th century. They didn't just speculate about this image of little balls bouncing off one another, but instead they developed a fully worked-out kinetic theory of gases founded upon Newton's laws.
* ***The word 'kinetic' comes from the Greek for motion, and according to Maxwell and Boltzmann's theory the pressure that a gas applies to the walls of its container is explained in terms of action and reaction (Newton's third law again) – each atom or molecule collides with the wall and bounces off, giving a push to the wall as it does so. This happens time and again, as the atoms rebound off each other and bounce back to hit the walls again.***

KEY WORDS

ATOM:
the smallest unit of a chemical element that can take part in a chemical reaction

MOLECULE:
two or more atoms of the same element or different elements held together by their chemical attraction

KINETIC THEORY:
theory describing the behaviour of matter in terms of the movement of its component atoms and molecules

MOLECULES IN MOTION

heating a solid breaks the bonds holding the molecules together

* A key feature of the kinetic theory is that it explains heat simply in terms of the motion of the molecules involved. If you heat up a container full of gas, the molecules move faster - so they give a bigger kick to the walls of the container each time they hit them, and the pressure increases. All of this was described mathematically, using equations (based on Newton's laws) that made it possible to calculate, for example, how much the temperature of a container full of gas would go up if it was heated by a particular amount.

KEY WORDS

THERMODYNAMICS: the branch of physics that deals with heat and motion (especially the way heat is transformed into other forms of energy)

SOLID TO LIQUID

* ***The kinetic theory also explains the differences between solids, liquids and gases.*** In a solid, the atoms and molecules are held together – we now know, by electric forces – but jiggle about slightly as if they were running on the spot. This is a bit like a restless theatre audience shifting in their seats during a dull play.

* When the solid is heated, the molecules jiggle about more and more

(which is why the solid expands), until they have generated enough kinetic energy (energy arising from motion) to break the bonds that hold them in place and are able to slide past one another relatively freely. ***The solid has now become a liquid.***

LIQUID TO GAS

* In a liquid, the molecules are still more or less in contact with one another, but constantly brush past each other. You might make an analogy with the jostling crowd of theatre-goers streaming out of the auditorium after the show.

gas molecules have enough energy to move freely past each other

* Carry on heating the liquid, and at a critical temperature the molecules will have so much energy that they fly freely past one another and can bounce off each other, ricocheting wildly, like balls in a crazy pinball machine. ***The liquid has now become a gas.***

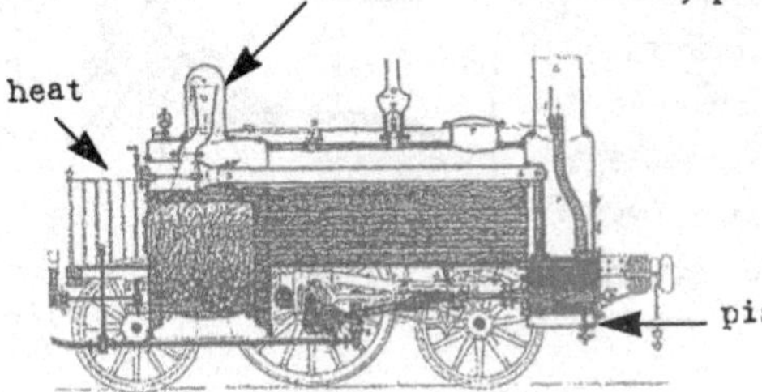

Piston power

If you imagine not a fixed container of gas but a cylinder fitted with a piston, you can see how the flying molecules in the gas will push the piston outwards. If the piston is held in place by a force pushing inwards, the hotter the gas inside the more force you will have to apply to the piston. This classic example of Newton's laws at work relates directly to the branch of science known as THERMODYNAMICS (the study of heat and motion). Thermodynamics was hugely important in the 19th century, because at the heart of the Industrial Revolution were steam engines – which were driven by pistons.

I hope I don't hit anything

SEQUENCE AND CONSEQUENCE

*** Something peculiar happens when you deal with large numbers of atoms and molecules. Although every collision between those individual molecules happens in accordance with Newton's laws, the interactions of all the molecules, taken as a whole, follow what we recognize as chronological time. It doesn't look peculiar, because it is what we are used to in everyday life - but in terms of Newtonian physics it really is very strange indeed.**

the flow of energy from hot objects to cool ones is a fundamental principle of the Universe

THE ARROW OF TIME

There's something curious about many of the experiments described so far. ***Newton's laws of motion do not take any account of the direction of the flow of time.*** It may ***seem*** as if there's an 'arrow of time' involved in Newtonian mechanics, because we can talk about some events occurring 'before' or 'after' others. But think about the simplest Newtonian interaction, when two billiard balls (or two atoms) move towards one another, collide and move apart. If you reversed the whole process, the backwards-in-time collision would still conform to Newton's laws of physics. ***Indeed, if you made a movie of such a collision and ran it backwards through a cine projector, the audience wouldn't suspect there was anything wrong.***

HALFWAY IN OR HALFWAY OUT?

* Think about that piston with the cylinder full of hot gas. As the gas pushes the piston, it moves it further and further out of the cylinder. This takes energy away from the molecules of the gas, so they move more slowly – they cool down. This is a fundamental feature of the Universe: heat flows naturally from a hot object to a cool one. To restore heat to the gas in the cylinder you would have to push the piston in, using energy to do so.

* If you saw a photograph of the piston pushed deep into the cylinder and another showing it much further out, you would know straight away which photo was taken first. When there are lots of molecules and atoms involved, nature has an inbuilt arrow of time.

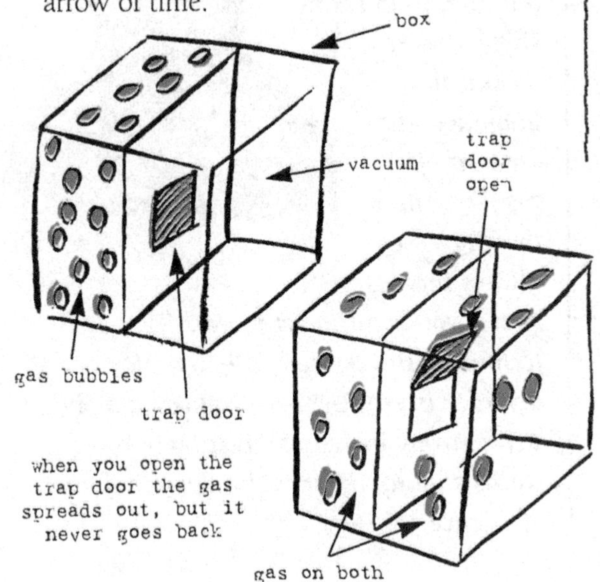

Half full or half empty?

Instead of a smoothly sliding piston, imagine a box divided into two halves by a wall, with gas in one side and a vacuum in the other. If you open a trap door in the dividing wall, the gas will spread so that it fills both halves of the box evenly (and it will cool down as it does so). No matter how long you wait, the gas will never, of its own accord, all move back into one half of the box. Again, if you saw a photograph of the box with all the gas in one half, and another photo showing the gas evenly spread through both halves of the box, you would know which photo was taken first. ***Nobody fully understands how the arrow of time emerges when interactions that individually take no notice of it are put together, but it is a fundamental feature of the physical world.***

KEY WORDS

DISORDER:
In thermodynamics, disorder doesn't just mean a mess, but a lack of pattern. A black-and-white chessboard has order. The same amount of paint making the board a uniform grey is disordered.

ENTROPY:
A measure of the amount of disorder in a system being studied, or in the entire Universe. The entropy of the Universe always increases.

THE FIRST LAW OF THERMODYNAMICS

This is in effect a preamble to the second law. It states that heat and work are two facets of the same thing, energy, and that the total amount of energy in a closed system stays the same.

THERMODYNAMICS

*** This business about the arrow of time and about heat always flowing from a hotter object to a cooler one is part of a law that is regarded as the most fundamental law in the whole of physics - the second law of thermodynamics.**

THE SECOND LAW OF THERMODYNAMICS

* The second law was established by the work of **William Thompson, 1st Baron Kelvin** (1824–1907), in England, and **Rudolf Clausius** (1822–88), in Germany, early in the 1850s. ***It can be summarized in three words: 'things wear out'. Or, to put it in slightly more technical language, the amount of*** DISORDER ***in the Universe always increases. And if you want to get more technical still, the scientific term for disorder is*** ENTROPY ***– so you can simply say 'entropy increases'. Just these two words sum up the most fundamental law of science.***

Lord Kelvin

INCREASING DISORDER

* The classic example of disorder (or entropy) increasing in this way is when you put an ice cube in a glass of water and watch it melt. The water with the ice floating in it has a kind of structure, a pattern. But when the ice cube melts (an example of heat flowing from the hotter object into the cooler object), there is just a featureless, amorphous, uniform blob of water. ***And again, the arrow of time appears – you often see ice cubes melting in glasses of water, but you never see a glass of water in which ice cubes appear spontaneously while the rest of the water warms up, even though that would not require any input of energy and so would not violate the first law of thermodynamics.***

as the ice melts, order is replaced by disorder

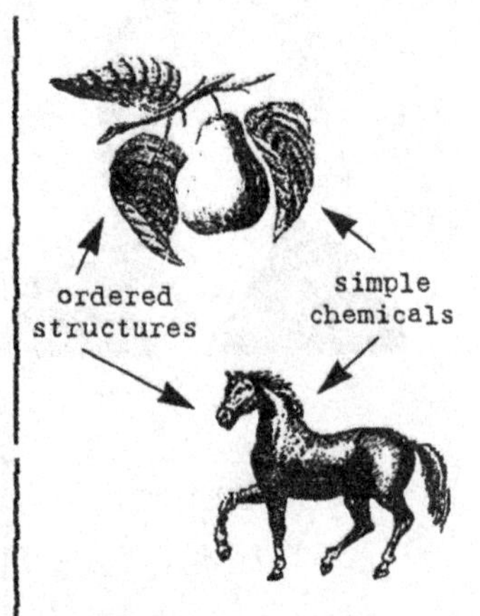

More or less entropy?

One thing that seems to violate the second law of thermodynamics is life itself. Plants and animals are very complicated ordered structures, built out of simple chemicals, that create order (thereby decreasing entropy) on a local scale. They are only able to do this with the aid of a large input of energy, which comes, ultimately, from sunlight. But the amount of order created by life on Earth in this way is more than compensated for by the amount of disorder (entropy) being created inside the Sun – by the processes that release energy in the form of sunlight. ***In the Universe at large, entropy always increases.***

Thomas Young

NEW LIGHT ON LIGHT

* As we shall see, the new physics offers at least one way of explaining problems such as entropy and where the arrow of time comes from. But before we get to grips with them, there's an important piece of old physics to consider - the physics of light.

THOMAS YOUNG (1773-1829)

A child prodigy, Young could read at the age of two, absorbed Latin and Greek as a child, and mastered several Middle Eastern languages before his teens. He had read and understood all Newton's work before he was 17. After qualifying as a doctor (in 1796), he became interested in optics through work on the human eye. As a result, Young carried out a series of experiments involving sound and light, and in 1800 published a book proposing (among other things) that light travels as a wave. He was also fascinated by Egyptology, and was instrumental in deciphering the Rosetta Stone.

Young spent his childhood reading clever books

A KEY CONCEPT

* The behaviour of light proved to be the key to the two great revolutions that swept through physics in the first decades of the 20th century – the quantum revolution and the relativity revolution. Ironically, though, these two breakthroughs occurred just after the theory of light had been put on what seemed to be a secure footing by the physicists of the 19th century – and by two of them in particular, **Michael Faraday** and **James Clerk Maxwell**.

WAVES, NOT CANNONBALLS

★ *Isaac Newton had had the idea that light is like a stream of tiny cannonballs, flying through space and bouncing off things.* This tied in with his laws of motion, so it was a natural model for him to adopt.

Augustin Fresnel invented a special lens for lighthouses

★ *Then at the beginning of the 19th century experiments by* Thomas Young *in England and* Augustin Fresnel *in France showed that light actually moves through space (or any transparent medium) in the form of a wave.* The clearest proof of this is a famous experiment used by Young, known as 'Young's double-slit experiment' or 'the experiment with two holes'. *It will be very important when we come to the new physics, so it is worth spelling out in detail what Young discovered.*

hmmm.. I wonder what light really is

Politics and optics

A civil engineer, Augustin Fresnel (1788–1827) became head of the public works department in Paris under Napoleon. He was also interested in optics and invented a special lens for lighthouses. When Napoleon was exiled to Elba, Fresnel supported the restoration of the monarchy, thus showing a good eye for the main chance. Alas for Fresnel, Napoleon came back, and he was placed under house arrest in Normandy, where he developed his wave theory of light. However, Waterloo brought Fresnel back into the open and he went back to engineering.

THE EXPERIMENT WITH TWO HOLES

all you need is some cardboard and a torch

* If you take a bright light and shine it on a piece of cardboard with a tiny hole in it, the light passes through the hole and spreads out on the other side. Now, you put a second piece of cardboard with two holes (tiny pinholes) in it in the path of the light spreading out from the first hole. The light spreads out from both of the holes in the second card. Finally, you put a third piece of cardboard in the path of the light spreading out from the two holes, and look at the pattern of light and shade that is made on this final screen (of course, you have to do this in a darkened room, in order to see the pattern at all). You get a pattern of alternating bright and dark bands (light and shade) - which can be explained if the light is travelling in the form of a wave, very much like ripples on a pond.

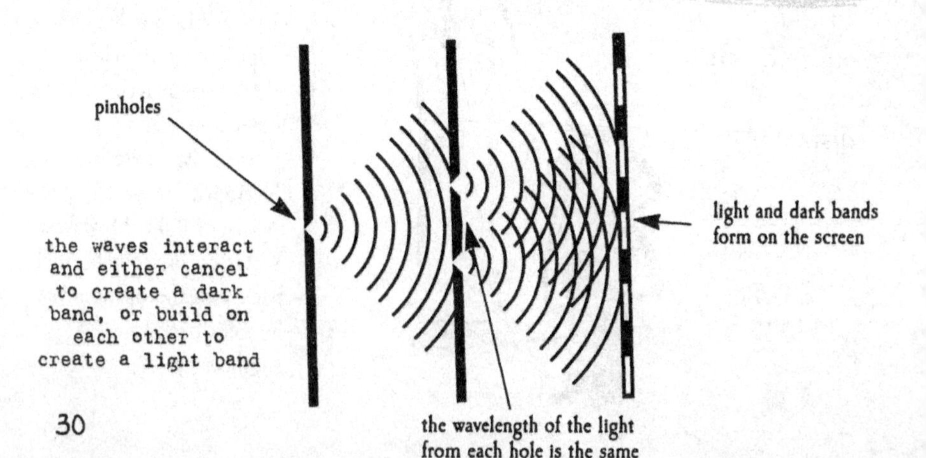

MAKING RIPPLES

* The waves from each of the two holes in the intermediate piece of cardboard start out in step with one another, because they come from the same single hole in the first piece of cardboard. They spread out like ripples on a pond produced by dropping two stones in at the same time, and they interfere with one another to make a more complicated ripple pattern.

light waves interfere with each other like ripples on a pond

PEAKS AND TROUGHS

* Where the waves overlap, in some places the peaks in the waves from each set of ripples coincide, so you get an extra high peak – a bright stripe on the far screen. In some places, the peak of one wave coincides with the trough of the other wave, so they cancel each other out and there is no light on the far screen – a dark stripe. And if two troughs coincide, that also produces a bright stripe, because the waves are adding together, even though they are adding in the opposite direction.

THE DOUBLE-SLIT EXPERIMENT

Young's original version of the so-called 'experiment with two holes' used narrow slits, cut with a razor, in the screens – which is how the experiment got its original name. With parallel slits, instead of pinholes, the pattern of light and shade produced on the final screen is simply a set of parallel stripes of light and shade, a distinctive interference pattern. By measuring the distance between the stripes in the pattern of light and dark on the final screen, it is possible to work out the wavelengths of the waves involved.

stop making it ripple it's making me sick

electricity and magnetism are a single force

FARADAY AND ELECTROMAGNETISM

* We all have some idea of the nature of electricity and magnetism from practical experience - but the experiments carried out by Michael Faraday, beginning in the 1820s, demonstrated that electricity and magnetism are actually a single force (electromagnetism) that shows two different facets to the world, depending on which way you look at it.

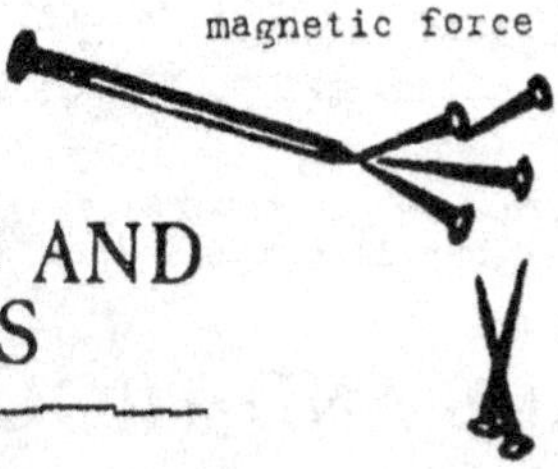

magnetic force

Faraday identified electromagnetism

The field of force

Faraday was not content with discovering the links between electricity and magnetism and showing how they could be put to practical use. He tried to explain how the effects worked, and in doing so invented the scientific concept of THE FIELD OF FORCE (often referred to simply as 'the field').

MOTORS AND MAGNETS

* *A moving electric charge (an electric current) produces a magnetic field – which is the reason why when electricity flows along a wire, the needle of an ordinary magnetic compass placed near the wire is deflected.* In fact, this is the principle of the electric motor – in which varying electric currents moving along wires make magnets with drive shafts attached to them spin round in circles.

MAGNETS AND DYNAMOS

Michael Faraday

* ***A moving magnet, on the other hand, makes an electric current flow in a nearby conductor.*** So if you wave a magnet about near a piece of wire, electricity flows in the wire. This is the principle behind the electric generator, or dynamo, in which magnets being whirled round mechanically on the end of a drive shaft make electric currents flow along wires. ***A dynamo is, in effect, an electric motor in reverse.***

wave that quicker I can't see with this going on and off

waving a magnet near a wire creates an electric current

MICHAEL FARADAY (1791–1867)

A self-taught genius, Faraday was the son of a blacksmith and only received a rudimentary education. While apprenticed to a bookbinder, he became fascinated by science, thanks to reading entries in a copy of the *Encyclopaedia Britannica* that he was working on. Through evening classes and wider reading he became more and more knowledgeable, then got a job as assistant to Humphry Davy at the Royal Institution, where he eventually succeeded Davy as Director. In addition to his work on electromagnetism and field theory, Faraday made important contributions to chemistry and served as a scientific adviser to the government. A member of a strict religious sect, he was extremely modest, declining a knighthood and the chance to be President of the Royal Society.

THE FORCE AND THE FIELD

★ The best way to get to grips with what is meant by a field of force is the same way that Faraday himself got hold of the idea - by means of the following experiment.

KEY WORDS

LINES OF FORCE:
An imaginary line indicating the direction of a force associated with a field. Think of it as a stretched elastic band.

FIELD:
The region in which one object exerts a force on other objects. There is a gravitational field associated with the mass of an object, an electric field associated with electric charge, and so on.

AN INTRIGUING WEB

★ Take an ordinary bar magnet and place a piece of paper on top of the magnet, then sprinkle iron filings on top of the paper. If you tap the paper gently, the filings arrange themselves in a series of curving lines, or arcs, linking the north and south magnetic poles of the magnet. ***These lines represent the field of force of the magnet, a kind of (usually) invisible spider's web stretching out all around the magnet. Faraday deduced that electric charges, as well as magnets, have their own field of force.***

★ Dynamo and electric motor effects result when the LINES OF FORCE cut across each other, making a complex web of interactions in the FIELD.

you can see the lines of force if you hold a magnet underneath a piece of paper with some iron filings sprinkled on it

For example, it is because magnetic lines of force are moving through the wire that the electricity is forced to flow in a dynamo, even though the moving magnet itself never touches the wire.

FARADAY'S FIELD THEORY

* Having worked out the nature of electromagnetism, Faraday took his thinking a step further. He suggested that the field idea ought to apply to the only other force known to 19th-century science, gravity. ***He pictured the Sun, for example, sitting in a web of force lines stretching out across space, and holding the planets – including the Earth – in its grip.***

Responding to the field

Let's see how Faraday's field theory applies to the Earth. ***All that matters is the nature of the field of force at the point in space where the Earth happens to be.*** Suppose you could remove the Earth from its orbit around the Sun. If, once the field had settled down, you were able to drop the Earth back into its orbit (at the right speed and so on), it would immediately feel the force and follow its orbit again. It wouldn't fly off into space before the Sun noticed it was there. Nor would it have to wait for a signal from the Sun instructing it how to move. The field is already there, in space, providing information about the existence of the Sun. All the Earth has to do is to respond to the field at the point in space where the Earth happens to be.

the Earth is held in its orbit by the Sun's field of force

AHEAD OF HIS TIME

*** Faraday's ultimate extension of his field theory was heady stuff, which his Victorian contemporaries didn't really understand. In fact, with these speculations he was nearly 100 years ahead of his time. In contrast, his ideas about electric and magnetic fields spreading out from their sources were taken up almost immediately.**

THE FIELD ITSELF

In the 1840s Faraday took his thinking further still. ***He suggested that – instead of thinking of magnets as the source of the magnetic field, electric charges as the source of the electric field, and massive objects like the Sun (or an apple, or the Earth) as the source of the gravitational field – the field itself was what really mattered.*** The Universe, he concluded, is full of fields of one kind or another. And things like magnets, electric charges and lumps of matter are simply places where the appropriate fields get knotted up in certain ways.

Faraday was nearly 100 years ahead of his time

FARADAY AND MAXWELL

* The person who took up these ideas was James Clerk Maxwell, who turned them into a complete, worked out mathematical theory of electromagnetism (published in 1864), and along the way, almost accidentally, explained the nature of light – or so it seemed at the time.

* It was Faraday who first suggested that light might be produced by some sort of vibration in the lines of force associated with charged particles and

Maxwell developed a colour theory of light still used in TV design

magnets – with the lines vibrating or twanging, rather like plucked violin strings, and with light being transmitted by waves running along the lines of force, outward from the source. But, despite all his success as a practical physicist and experimenter, and his remarkable insight into the nature of fields, Faraday was no good at mathematics and he was never able to convert this vague suggestion about the nature of light into a comprehensive theory.

Viol

James Clerk Maxwell (1831–79)

When, at the age of 10, Maxwell went to school in Edinburgh, his country-bumpkin ways earned him the nickname 'Dafty'. Nevertheless, he became a professor in Aberdeen, then in 1860 moved to King's College, London. Following his father's death, in 1865, Maxwell retired to look after the family land. In 1871 he was persuaded to head the new Cavendish Laboratory in Cambridge. He had just established it as a scientific centre of excellence when he died, at the age of 48. In addition to his work on electromagnetism, he proved that the rings of Saturn had to be made up of myriads of tiny moons, developed a colour theory of light still used in TV design, and made major contributions to thermodynamics and statistical mechanics.

James Clerk Maxwell

Maxwell's equations tell you all you need to know

MAXWELL'S EQUATIONS

* For all practical purposes, the set of equations that Maxwell came up with is valid for every electric or magnetic phenomenon that you are likely to come into direct contact with in everyday life at the beginning of the 21st century, except perhaps for the occasional laser beam. They describe all of the aspects of electromagnetism that are covered by the description 'classical' physics - meaning everything that does not involve quantum effects.

Measuring the speed of light

The speed of light was first measured by Danish astronomer **Olaus Roemer** (1644–1710) from an analysis of the eclipses of the moons of Jupiter in 1675. Ground-based measurements became accurate in the 1860s, when French physicist **Léon Foucault** (1819–68) developed a technique in which a beam of light was bounced between mirrors, giving a long enough path for its 'flight time' to be measured.

FOUR FUNDAMENTAL FORMULAE

* There are just four of these equations. Known simply as Maxwell's equations, together they describe all classical electric and magnetic phenomena – the dynamo effect, how electric motors work, why compass needles point north (so long as they are not close to a wire carrying an electric current, or to a permanent magnet), how big the force between two electric charges of a certain size a certain distance apart is, and much more besides.

compass

* *Every problem involving electricity and*

magnetism known at the time could be explained using Maxwell's equations.

ANSWERS TO EVERYTHING

★ *This was a huge and wonderful leap forward for science, which put Maxwell almost on a par with Newton.* After all, Newton's laws described the workings of everything in the known world of physics of the mid-19th century except electromagnetism – and Maxwell's

equations explained everything there was to explain (at that time) about electromagnetism. Together, Newton and Maxwell explained everything known to physical science.

AN UNEXPECTED BY-PRODUCT

The greatest and most wonderful thing about Maxwell's equations was the way that a description of light fell out of them without being asked for. Among the many things that the equations describe are the properties of electromagnetic waves – ripples in the electromagnetic fields. And the most important property of these waves is the speed with which they travel. A couple of numbers, specifying the strength of the electric and magnetic parts of the interaction, had to be put into the equations from experimental observations and measurements. ***But once those numbers were fed into them, the equations specified, uniquely and precisely, the speed with which electromagnetic waves must move. That speed turned out to be exactly the same as the speed of light, which by the 1860s had been measured very accurately.***

WAVES AND WAVELENGTHS

*** What distinguishes the different varieties of electromagnetic wave from one another is the wavelength of the radiation - just as different musical notes (which are all sound waves) are distinguishable by their different wavelengths.**

as you move the rope up and down, waves travel along it

Electromagnetic Waves

Visible light, radio waves, the infrared heat you feel as the warmth radiating from a radiator, the microwaves that cook the food in your microwave oven, X-rays, and even bursts of gamma radiation from distant galaxies are all varieties of electromagnetic radiation. All of them are described by Maxwell's equations.

LIKE A JIGGLING ROPE

* ***The best way to get a mental image of how electromagnetic waves travel through space is to picture waves rippling along a tightly stretched rope.*** If you tie one end of a rope to a fence post and hold the other end in your hand, you can make ripples run along the rope by jiggling the end you are holding up and down. The energy that makes the ripples move along the rope comes from the work you are doing by moving the end you are holding. For

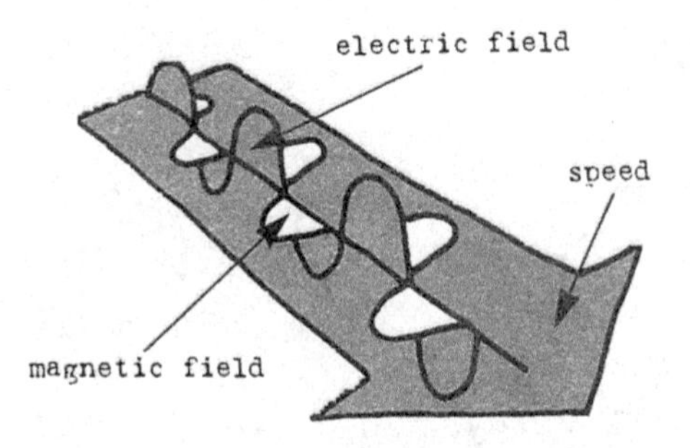

an electromagnetic wave is really two waves at right angles to each other - one is the electric field and one is the magnetic field

variety, as a change from making the ripples move vertically (up and down), you can make them move horizontally (to and fro).

DOUBLE ACT

* The energy that makes electromagnetic waves move through space is provided by electric charges (or magnets) being jiggled to and fro, in much the same way that the energy to make the ripples in the rope comes from jiggling it horizontally.

* ***But electromagnetic waves are more complicated, because they are composed of two waves running along together and sustaining each other. In fact, as long as the waves are at right angles to each other, the whole pattern can be twisted and oriented in any direction – but it is easier to think of them as vertical and horizontal waves.***

thank goodness for electromagnetic radiation

microwave ovens use electromagnetic radiation

ETERNAL TWOSOME

Suppose the vertical wave is an electric wave, produced by jiggling an electric charge up and down. It is a changing electric field, moving through space. But as Faraday discovered, a changing ***electric field*** produces a ***magnetic field***, at right angles to the electric field. So the vertical electric wave is accompanied by a horizontal ***magnetic wave***, moving through space. And as Faraday also discovered, a changing ***magnetic field*** produces an ***electric field***, at right angles to the magnetic field. So the horizontal ***magnetic wave*** is accompanied by a vertical ***electric wave***, moving through space. And so on. ***It doesn't matter whether you start with a jiggling magnet or a jiggling electric charge. The two changing fields sustain each other, and you can't have one without the other.***

PARTICLES IN A BOX

★ If you have a box of a certain size, filled with gas at a certain temperature and pressure, no matter what kind of gas it is there will always be the same number of particles (that is, atoms or molecules) bouncing around in the box. How do we know? Because they produce the same pressure on the walls of their container.

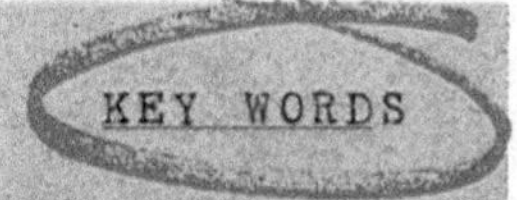

ELECTRON: negative particle found in the outer part of an atom (it carries one unit of electricity)

Amadeo Avogadro

JUST THE RIGHT SIZE

★ For a gas at 0°C and 1 standard atmosphere of pressure (the usual example used), you can make the box just the right size to hold a mass of gas equal to the molecular weight of that gas expressed in grams. The volume of the box would be 22.4 litres. Atomic and molecular weights are measured in units based on the mass of a hydrogen atom, which counts as 1. There are two hydrogen atoms in each molecule of hydrogen, so the molecular weight of hydrogen is 2, and the box would hold 2 grams of hydrogen. Similarly, each oxygen atom has an atomic weight of 16 (16 times the atomic weight of hydrogen) and there are two atoms in each molecule of oxygen, so the filled box would hold 32 grams of oxygen.

Making waves

The energy that makes electromagnetic waves has to be generated in the first place, by moving electric charges or magnets to and fro. The most common way we do this is by means of electric currents – which are just electric charges, consisting of huge numbers of ELECTRONS – flowing along wires. Discovered in the 1890s, electrons play an important role in the new physics – though they seemed to fit neatly into the world of Newtonian mechanics and Maxwell's equations.

the Greeks and Chinese had a form of atomic theory

AVOGADRO'S NUMBER

* Whatever the gas used, the number of molecules in the box will always be the same – a number known as Avogadro's number, which is 6×10^{23} (a 6 followed by 23 zeros, or a hundred thousand billion billion).

* ***This huge number, which was worked out by the Italian physicist* Amadeo Avogadro *(1776–1856) in 1811, gives you some idea of just how small atoms and molecules are. Yet by the end of the 19th century physicists were beginning to divide atoms into their component parts. Atoms were not indestructible, after all.***

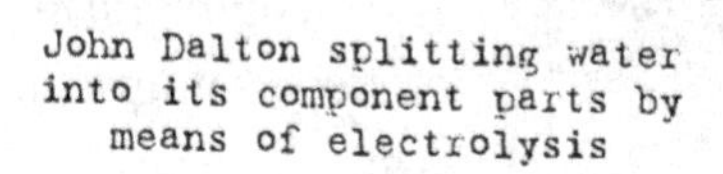

John Dalton splitting water into its component parts by means of electrolysis

ESSENTIAL PARTICLES

In ancient times both the Greeks and the Chinese had thought of the world as being composed of 'essential' particles, but the modern idea of atoms developed in the 17th century and later, from the work of scientists like **Robert Boyle** (1627–91) and **Christiaan Huygens** (1629–93). In 1738 **Daniel Bernoulli** (1700–82) suggested that a gas might be made up of tiny particles that bounced around within their container. But it was only at the start of the 19th century that **John Dalton** (1766–1844) proposed that an atom was the smallest unit of an element that could take part in a chemical reaction.

NOVEL EXPERIMENTS

*** In the middle of the 19th century physicists began to study the effect of electric currents on traces of gas left in glass tubes from which air had been almost completely evacuated - now possible thanks to the invention of the vacuum pump.**

physicists were interested in the effect of electricity on gases

J.J. THOMSON (1856-1940)

In 1897 Thomson demonstrated that electric currents consisted of streams of electrically charged particles (electrons). Although he devised and supervised the experiments, he never did them himself – he was so clumsy that his colleagues claimed the apparatus would break if he even looked at it. Head of the Cavendish Laboratory during its 'glory years' around the start of the 20th century, he received the Nobel Prize in 1905 and was knighted in 1908.

ILLUMINATING DISCOVERIES

* For these experiments, the current was made to flow between a positively charged plate at one end of the tube (the anode) and a negatively charged plate at the other end (the cathode). This resulted in a stream of negatively charged rays streaming away from the cathode. Logically enough, these were dubbed CATHODE RAYS.
In 1895 German physicist **Wilhelm Röntgen** (1845–1923) discovered X-RAYS – a previously unknown form of radiation, produced when cathode rays hit the glass wall of the tube. *X-rays proved to be a very energetic (short-wavelength) form of electromagnetic wave.*

J.J. Thomson

* Two years later, in 1897, a British physicist, **J.J. Thomson** (1856–1940), showed that cathode rays were quite different from X-rays – ***being a stream of tiny particles, much smaller than atoms, that carry a negative electric charge.*** These were soon given the name 'electrons', by the Dutch physicist **Hendrik Lorentz** (1853–1928).

TWO NAGGING QUESTIONS

* So, at the end of the 19th century, the scene was set for the new physics. Physicists had Newton's laws to describe the behaviour of matter, and Maxwell's equations to describe radiation. They had an emerging picture of the atom as a not-quite-indivisible particle, which could be described in terms of Newtonian mechanics to explain such things as the behaviour of a gas in a container, but which could be subdivided (somehow) by having negatively charged electrons (which themselves seemed to be Newtonian particles) chipped away from them, leaving a positively charged remnant (an ion), which also obeyed Newton's laws and Maxwell's equations. Everything in the garden looked rosy – except for two nagging questions about the behaviour of light.

Glowing achievement

Wilhelm Röntgen discovered X-rays while studying the way an electric current flowing through an evacuated glass tube makes the tube glow. He noticed that radiation given off by the tube was causing a nearby fluorescent screen to glow, too. He received the first Nobel Prize for Physics, in 1901.

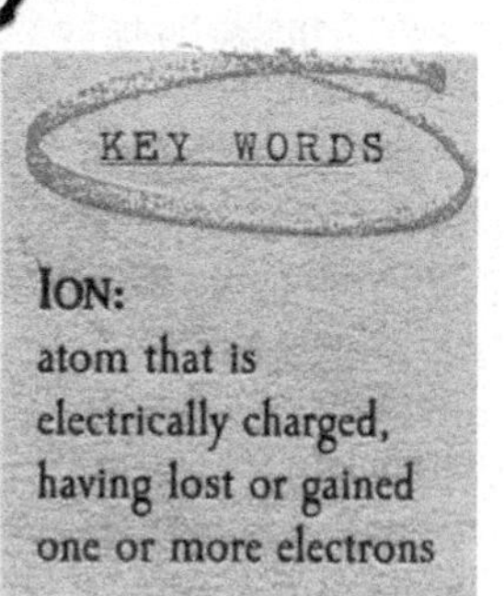

we have Röntgen to thank for X-ray spex!

KEY WORDS

ION:
atom that is electrically charged, having lost or gained one or more electrons

this physics is really new territory

Postprandial physics

Because he was believed to be erratic, Einstein found it impossible to obtain a university job, even as a junior assistant. But he kept on thinking about physics in his spare time – or rather, he thought about physics full time and only did enough other work to make ends meet. The story is that in the patent office he was so good at understanding the technicalities of the patent applications he was judging that he would zip through a day's work before lunch, then devote the afternoon to thinking about scientific problems – including the ones he resolved in his special theory of relativity.

BEYOND NEWTON

* By a neat coincidence, the new physics begins at the beginning of the 20th century, so it is now 100 years old. But it is still thought of as 'new' because it uses ideas and concepts that go beyond those of Newtonian physics. Even Maxwell's wave theory of light would have been entirely intelligible to Newton (although he might have been disappointed to find that his own 'corpuscular' theory of light had not been proved valid), but with the new physics he would have been entering unfamiliar territory.

new physics is nearly 100 years old

'LAZY DOG' – OR QUICK FOX?

* Albert Einstein *(1879–1955) worked out his first innovative theory – the special theory of relativity – in the first years of the 20th century, and published it in 1905.*

Einstein wouldn't work at anything unless it interested him

At the time he had no university appointment, but was working as a technical expert (second-class!) at the Swiss patent office, in Bern.

* Although he had graduated, in 1900, with reasonably good marks in his final examinations, as an undergraduate Einstein had gained a reputation for being – in the words of one of his tutors, Hermann Minkowski (see page 66) – 'a lazy dog' who wouldn't work at anything unless it interested him. In fact, having skipped most of the lectures, he only got his degree by cramming frantically for a few weeks, using lecture notes taken by one of his friends, Marcel Grossman.

I'm no lazy dog

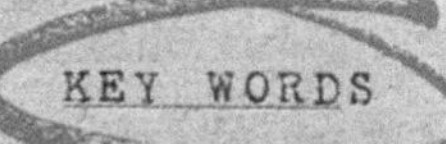

KEY WORDS

RELATIVITY: According to Einstein's theory, any 'observer' is entitled to consider that he or she is 'at rest'. All motion is therefore relative to the observer.

Albert Einstein

Enter Einstein

The first part of the new physics to be completed as a fully worked-out theory – and tested by experiments which proved that it was 'better' (that is, more complete) than Newtonian mechanics – was Einstein's theory of relativity. Or rather, his two theories of relativity.

A PUZZLING PARADOX

'but officer I was only doing 30'

* The thing that set Einstein thinking about relativity was a conflict between Newtonian physics and Maxwell's equations. Remember that one of the greatest triumphs of Maxwell's equations is that they automatically give the speed with which electromagnetic waves travel through space - a value identical to the speed of light, which never varies.

the speed with which anything passes you depends on how fast you are moving

THE SPEED OF LIGHT

The speed of light is usually denoted by the letter c (for 'constant'). It is almost exactly 300,000 kilometres per second.

THE NEWTONIAN VIEW

* ***In Newtonian mechanics, on the other hand, the speed with which anything passes you depends on how fast you are moving.*** If I stand by the side of a road and a car drives past me at 60 kilometres per hour, the car is regarded as moving at 60kph relative to the road. And relative to the road, I am stationary. Similarly, if I am in a car moving at 30kph relative to the

road and am overtaken by a car moving at 60kph relative to the road, then the second car is travelling at 30kph relative to me, not at 60kph.

THE MAXWELLIAN VIEW

* ***But for light, Maxwell's equations make no allowance for any effect of this kind.*** If I stand by the side of the road and measure the speed of light coming from the headlights of a car moving towards me at 60kph, according to Newtonian mechanics the speed I measure ought to be $c + 60$. But according to Maxwell's equations, the speed of light is c, pure and simple.

* It doesn't matter if the source of the light is moving towards you, or away from you, or round in circles, or standing still. Indeed, it doesn't matter if you are in a spaceship travelling at half the speed of light relative to the Sun and measure the speed of the light coming from the Sun. ***Maxwell's equations insist that the speed of light is always c.***

* And if there are two spaceships, one hurtling towards the Sun at half the speed of light and the other hurtling away from the Sun at half the speed of light, they will still each measure the speed of light from the Sun as c – as will anybody on any of the planets, although they orbit around the Sun at different speeds.

Trains, not spaceships

Einstein did not, of course, talk about spaceships. In fact, he liked to use examples involving railway trains hurtling along the tracks. But the underlying point was the same: Maxwell's equations do not allow for the possibility that velocities involving light add up in the way described by Newtonian mechanics – the 'common-sense' way, in which 2 + 2 = 4.

the speed of light is the same however the source is moving

Maxwell versus Newton

At the end of the 19th century the Newtonian view of the world, by then established for more than 200 years, was endowed with almost the authority of Holy Writ. In contrast, Maxwell's theory of light was a newcomer, only a few decades old. Maxwell himself had died as recently as 1879 and had not achieved the almost mythical status of Isaac Newton.

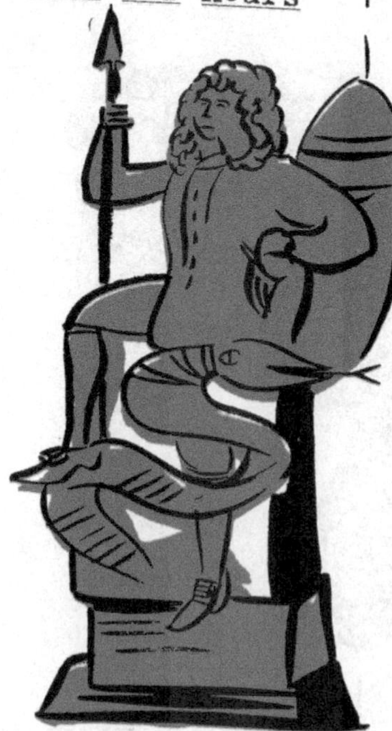

Newton was seen almost as God in the 19th century

CONFLICTING VIEWS

* The few people who were bothered by the apparent conflict between Newtonian and Maxwellian theory tended to assume that there was something wrong with Maxwell's equations. But Einstein's great quality was his refusal to take anything for granted.

people began to be bothered by the conflict between Maxwell and Newton

THINKING THE UNTHINKABLE

* The very things that made Einstein such a poor student – his independence, his questioning of received ideas, and ability to work stubbornly at problems that interested him – made him start to think the unthinkable.

* What if Maxwell were right and Newton were wrong? What would the world be like if the laws of mechanics really were set up in such a way that the speed of light is exactly the same for all observers, everywhere in the Universe, no matter how they are moving relative to each other, or relative to the source of the light?

EINSTEIN'S SPECIAL THEORY OF RELATIVITY

things moving in circles

★ ***At first, Einstein restricted himself to thinking about things moving at constant velocities relative to one another – at constant speeds in straight lines.*** He did not try to deal with accelerations, either straight-line accelerations or things moving in circles and other curved paths. ***This is the sense in which the theory he came up with in 1905 is 'special': the term means 'restricted', as in something being a special case of a more general phenomenon.*** But although his theory was restricted in this sense, it was nothing like anything that had been seen before.

★ Einstein found that in order for the speed of light to be the same for all observers, velocities could not add up in the way Newton thought they did (2 + 2 = 4). ***So he formulated a new law to describe the way velocities add up, which is described by a simple mathematical expression.***

$$V = \frac{(v_1 + v_2)}{(1 + (v_1 v_2)/c^2)}$$

important equation

NEWTON'S MATHS AND MAXWELL'S

If v_1 and v_2 are the velocities of two objects moving towards one another through space, in Newtonian terms they would be approaching at a combined speed V, with $V = v_1 + v_2$. But if we go by Maxwell's equations, then we have to use the expression

$$V = \frac{(v_1 + v_2)}{(1 + (v_1 v_2)/c^2)}$$

In other words, the Newtonian way of adding up the velocities has to be divided by a number equal to 1 plus the result of multiplying the two velocities together and dividing them by the square of the speed of light.

NEWTON DIVIDED BY ONE

* There are two important things about Einstein's equation. The first is that if the velocities involved are a lot smaller than the speed of light, by the time you do the dividing by the speed of light squared, that bit of the expression is tiny. So, in effect, you get the Newtonian 'answer' divided by 1.

Einstein's maths are hard to follow

DIY relativity

It's easy to use a pocket calculator to work out exactly how two velocities less than c add up according to Einsteinian mechanics. Try calculating the relative velocity for two spaceships that are approaching each other head on, each moving at three-quarters of the speed of light relative to an observer on Earth. You should find that the people on board each of the spaceships will measure the velocity of the other spaceship as $0.96c$.

AN IMPERCEPTIBLE DIFFERENCE

* This means that at slow speeds (slow compared with the speed of light) the combined velocities are so nearly in line with Newtonian mechanics that you cannot measure the difference. ***In other words, for everyday things like cars driving down roads, or trains hurtling along railway tracks, Einstein's way of adding up velocities gives exactly the same answers as Newton's.***

I'm going as fast as I can

THE SPEED OF LIGHT PLUS THE SPEED OF LIGHT

★ *The other important feature of the Einsteinian way of adding up velocities is that if you look carefully at the new equation you will see that you can never add up two velocities that are smaller than the speed of light (less than c) to give a relative velocity larger than the speed of light.*

for speeds less than the speed of light, Newton and Einstein give the same answer

★ If the two objects we are interested in – spaceships, trains or whatever – are moving towards one another at the speed of light, then both v_1 and v_2 are equal to c. So the top of the fraction becomes $2c$. But v_1v_2 (which simply means $v_1 \times v_2$) becomes $c \times c$, which is c^2, so the bottom of the fraction becomes $(1 + c^2/c^2)$, which is 1 + 1, which any child knows makes 2. So the 2s cancel out, and the relative velocity V is equal to c. In Einsteinian mechanics, $c + c = c$.

special case

A SPECIAL CASE

Newtonian mechanics isn't 'wrong'. It is in effect the version of Einsteinian mechanics that applies for velocities much smaller than the speed of light – a 'special case' of the special theory of relativity. ***To put it another way, Newtonian mechanics, in all its glory, is entirely contained within Einstein's description of the world.***

it's child's play

STANDING STILL IN SPACE

Because relative motion is so important, we have to be clear exactly what we are talking about. ***In all this discussion, we imagine an observer standing still in space, while various objects whiz past at high speeds relative to the observer.*** To illustrate this point, Einstein wrote about an observer standing on the platform of a station, with high-speed trains whistling along the tracks.

EXACTLY THE SAME?

*** But does the new way of adding up velocities really always give the same speed for light itself? Yes, subject to the following provisos. If either v_1 or v_2 is equal to c, then whatever the other velocity you put into the calculation, the answer you get is always c - provided that the other speed you put in is less than c.**

TWO PROVISOS

* If you try this on your calculator, though, you'll probably need to put a lot of digits in after the decimal point to make it work, because pocket calculators are only approximate guides to reality, whereas nature has as many decimal places as she needs to do the trick.

various objects whiz past at high speeds relative to the observer

* *The speed of light through space really is exactly the same for all observers. However, it is true that light moves slightly more slowly when passing through something like*

glass, or water, or even air. So, strictly speaking, this discussion applies only to the speed of light travelling through empty space – through a vacuum.

IN THE FAST LANE

* On its own, Einstein's version of the rule for adding up velocities wouldn't be very sensational. ***But it is just one part of a comprehensive theory that describes what happens to objects travelling at close to the speed of light.***

* Of course, it also describes what happens to slow-moving objects – but for them, as we have seen, Einstein's equations give exactly the same answers as Newton's. ***In the context of the special theory of relativity, the fast-moving objects are the ones that are interesting.***

Relativistic effects

Einstein's equations tell us that the mass of a moving object (as measured by the observer) increases, while its length (again, as measured by the observer) contracts. Strangest of all, they also tell us that time measured on a moving clock runs more slowly than time measured by a clock stationary relative to the observer's clock (see page 58).

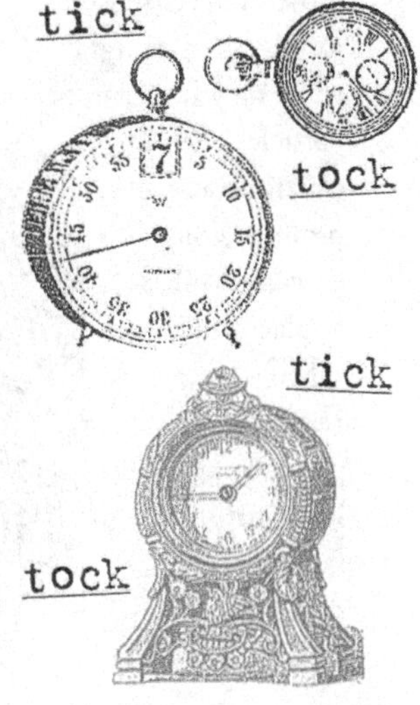

SIMPLY AND ELEGANTLY EXPRESSED

*** As with the addition of velocities, Einstein's relativistic effects are described by surprisingly simple mathematical expressions. Indeed, they are all described by the same conversion factor.**

THE CONVERSION FACTOR

* If v is the velocity of the object relative to the observer, then the appropriate expression that comes into the calculation is the square root of $(1-v^2/c^2)$, which is written as $\sqrt{(1-v^2/c^2)}$. The mass of an object moving at velocity v is equal to the mass of the same object when stationary (relative to the observer) divided by this factor, while the length of the moving object is equal to its length when stationary (always referred to as the 'rest length') multiplied by this same factor.

'the maths is very simple'

**** The bigger v is (up to the speed of light), the smaller the factor (which is always less than 1). Anything it multiplies is therefore smaller, while anything***

Relativity in the lab

Ideas about the effects of relativity are difficult to accept, but they have all been tested by experiments. Physicists play with things that move at speeds close to the speed of light when they fire charged particles such as electrons and protons along evacuated tubes at places like CERN (the European particle-physics laboratory) or Fermilab (the American counterpart).

divided by it is bigger. And the time between ticks of a moving clock is equal to the time between ticks of the same clock when at rest, again divided by the factor $\sqrt{(1-v^2/c^2)}$.

TAKEN TO EXTREMES

* For low velocities, v is much less than c. So v^2/c^2 is very small, and the factor reduces to the square root of 1, which is itself 1. ***There is therefore no noticeable difference from the Newtonian world.***
* At the other extreme, if v is equal to c, the factor becomes zero (1 minus 1). Anything multiplied by zero is zero, and anything divided by zero is infinitely big. ***Consequently, as things approach the speed of light they become heavier and heavier, without any limit, and at the same time become shorter and shorter, shrinking away to nothing. With a moving clock, the ticks stretch out more and more as its speed approaches the speed of light. Anything moving at the speed of light – such as light itself – does not notice the passage of time at all.***

TESTING RELATIVITY

It is possible by monitoring how beams of particles respond to magnetic fields, to measure the extent to which the masses of the particles are changed by the effects of relativity. Even the strangest effect of all, the so-called TIME-DILATION EFFECT, has been measured (see page 58).

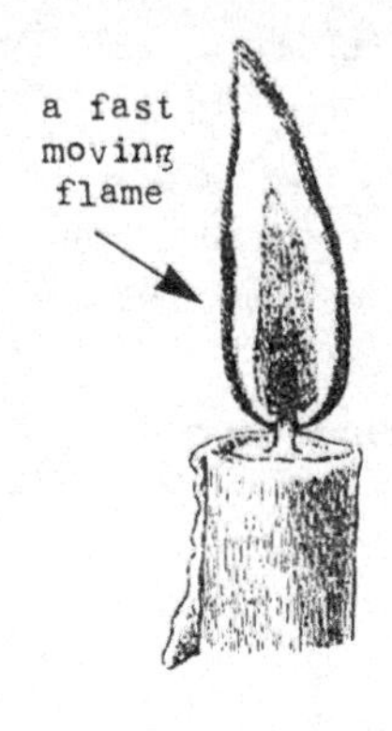

TIME DILATION

* It's worth explaining how time dilation can be monitored in the laboratory. I'll describe a slightly idealized hypothetical version of the experiment. Although no one has conducted an experiment quite this way, equivalent tests have been carried out in a slightly more complicated experimental set-up and have confirmed the accuracy of Einstein's equations.

nature provides us with a variety of different 'clocks'.

KEY WORDS

TIME DILATION: the 'stretching out' of time (the time recorded on a moving clock runs slow compared to the clock of a stationary observer)

LIFETIME: the length of time unstable particles 'live' in a particular form, before spontaneously turning into something else

TIMED DECAY

* Nature provides us with a variety of different 'clocks'. ***For example, when particles are produced in nuclear interactions they live for a certain length of time before they are spontaneously converted into other forms (in physicists' jargon, they 'decay'). Measurements in the lab tell us how long, when it is stationary relative to the observer, each kind of particle lives before it decays.***

* Imagine such a particle being shot along a perfectly straight tube, exactly 100 metres (10^4cm) long. For the sake of this example, let's choose a variety of particle that is known to live for exactly one-millionth (10^{-7}) of a second, which is

typical of the kind of LIFETIME these unstable particles have. Even if it were travelling at the speed of light (which is 3×10^{10} centimetres per second in the same notation), without time dilation such a particle could only travel a distance of 30 metres before expiring, because $(3 \times 10^{10}) \times 10^{-7} = 3 \times 10^{3}$cm = 30m. In this particular experiment, it couldn't even get halfway to the other end.

RELATIVISTIC CORRECTION FACTOR

This is the number used in calculating how much a moving clock runs slow or a moving object shrinks, compared with the properties the same object has when at rest.

WHIZZING DOWN THE TUBE

* Now think about an identical particle travelling at twelve-thirteenths of the speed of light (again, an entirely realistic speed in this kind of experiment), and make allowance for time dilation.
* Without time dilation, the particle would get 30 x ($^{12}/_{13}$) metres down the tube – a distance of 27.7 metres. But (as you can easily verify, using the relativistic correction factor and your pocket calculator) at twelve-thirteenths of c, the lifetime of the particle is stretched by a factor of 2.6. Since it lives 2.6 times as long (according to stationary clocks in the laboratory), it can get 2.6 times as far down the tube – a distance of 72 metres.
* ***By measuring how far such particles travel in experiments of this kind, physicists have observed time dilation at work and have confirmed on very many occasions that it really does obey Einstein's equations.***

how slow is this moving clock running?

SO WHAT'S THE PARTICLE'S VIEW?

* But how do things look from the particle's point of view? It is a key tenet of Einstein's theory that (so long as we are dealing with constant velocities) the moving object is entitled to consider itself at rest. There's no reason for it to feel any of the effects we are talking about; it doesn't feel any heavier, or notice that it has shrunk, or detect anything wrong with its clocks. So how can it get further down the tube than it 'ought' to?

IT'S THE LAB THAT MOVES

* *According to Einstein, because the particle is entitled to regard itself as being at rest, it follows that the laboratory (and everything else on Earth) is hurtling past it at twelve-thirteenths of the speed of light.*

* Now, as already mentioned, the particle only lives for a millionth of a second. But the tube that is flying past it is 'shrunk' by its high velocity – and it is shrunk by exactly the same factor as we used in the time-dilation calculation, 2.6. So the

KEY WORDS

FRAME OF REFERENCE: the place measurements are made from – the 'point of view' of an observer particle still gets 2.6 times further towards the end of the tube than it would if there were no relativistic effects at work. Everything fits together and it all works perfectly, no matter how you are moving, ***provided you are moving at a constant velocity.***

FRAMES OF REFERENCE

* ***Things that are moving at a certain (constant) velocity are said to be in a certain*** FRAME OF REFERENCE, ***and any frame of reference can be chosen as the one you make measurements from.*** The observers in each frame of reference think their own clocks, rulers and so on are perfectly normal, and it is everyone else's clocks and rulers that are affected by relativistic effects. But when you compare notes with observers in other frames of reference, ***you always get consistent answers about how the Universe works – not necessarily the same answers, but consistent ones.***

DIFFERENT POINTS OF VIEW

Because observers moving at different velocities are in different frames of reference, they each have their own picture of the Universe and their own ideas about, for example, whose clocks are running slow. Observers in two different frames of reference won't even agree on the mass of an object moving past both of them in a third frame of reference. But they can agree to differ, because they can each calculate what mass the object would have if it were brought to rest in their own frame of reference and will each get the same answer to this calculation.

THAT MOVING CLOCK AGAIN

* There's one more example worth looking at in detail that helps us understand how and why a moving clock can run slow. People often argue about which kind of clock is most reliable, and you might perhaps be worried about how the change in mass of the moving clock would affect its timekeeping properties. But there is an ultimate clock that cannot be argued with, and that is light itself.

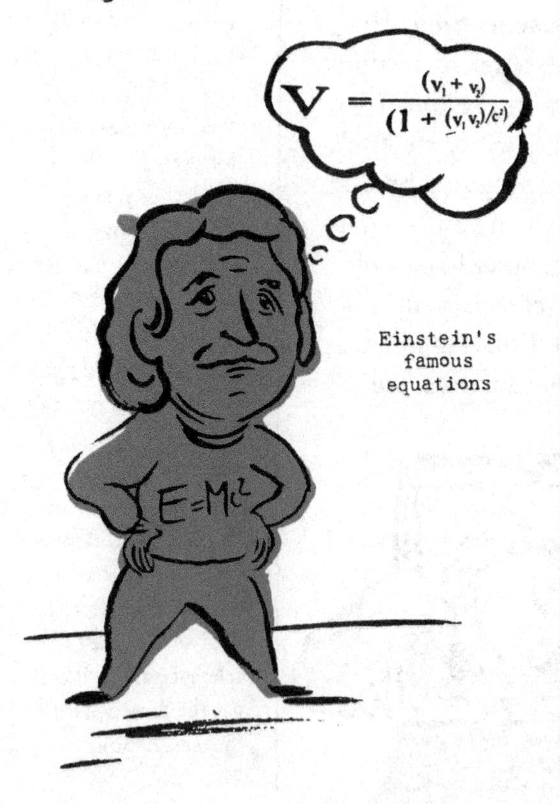

Einstein's famous equations

EINSTEIN'S FAMOUS EQUATION

This business about moving objects gaining mass explains where the most famous equation in all of science, $E = mc^2$, comes from. According to Newtonian physics, if you push an object you put energy in to make it move faster – it gains kinetic energy of motion simply by increasing its velocity (acceleration equals force divided by mass). ***But in Einsteinian physics, some of the energy you put in when you push an object goes into making it move faster – and some of the energy goes into making it heavier.*** This tells you that ***mass and energy are equivalent and interchangeable,*** and leads to the relation expressed in Einstein's famous equation by way of a calculation just a little too complicated to bother with right now.

the faster
you push,
the heavier
it gets

LIGHT TIME

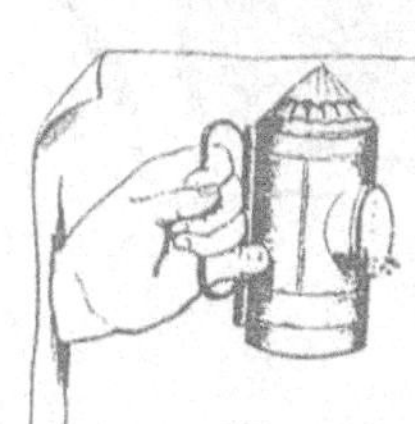

★ ***Since light always travels at the same speed, for all observers, it provides the ultimate measure of time.*** We can define one second as the time it takes for light to travel a certain distance (and this is, in fact, the way time is defined today, in terms of the properties of a particular wavelength of light emitted by caesium atoms).

★ ***There is no argument about this. In any frame of reference, you can choose light from any source in the Universe, measure how long it takes to cover a chosen distance, and work out how much time has passed.***

Guaranteed accurate

The ultimate clock would be a kind of light clock in which light bounces up and down between two perfectly shiny mirrors, situated a set distance (perhaps a metre) apart. Each tick of the clock would correspond to the time it took for a pulse of light to go from the top mirror down to the bottom mirror and back to the top mirror again, giving a steady beat for anyone in the same frame of reference as the clock.

ZIGZAG LIGHT

* How do things look to an observer in another frame of reference, watching the light clock just described move past at a constant velocity? We have to bear in mind that the light pulse is moving sideways as well as up and down. In the time it takes for the pulse to travel from the top mirror to the bottom mirror, the whole clock has moved sideways.

Pythagoras' theorem

The square of the hypotenuse of a right-angled triangle equals the sum of the squares of the other two sides.

c a 90° b

$a^2 + b^2 = c^2$

right-angled triangle

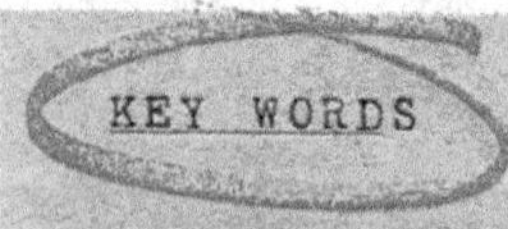

SPACETIME: the union of time with the three dimensions of space, making a four-dimensional whole

ENTER PYTHAGORAS

* To the observer in the second frame of reference, the light pulse flies along a diagonal path to the other mirror, and along an equivalent diagonal path back up to the top mirror. As this process repeats, the light zigzags up and down between the two mirrors.

* *Anyone who remembers learning at school about right-angled triangles and Pythagoras' theorem, will be aware that the diagonal path taken by the light beam is the hypotenuse of a right-angled triangle, and so*

Pythagoras

is longer than the vertical path between the mirrors. Since it is a longer path, the light must take longer to complete its journey. The moving clock must therefore run slow.

* The geometry of right-angled triangles tells us how much longer the path is, and therefore how much slower the moving clock is. The correction factor is, of course, Einstein's relativistic factor that we met before: $\sqrt{(1-v^2/c^2)}$.

* The faster the moving clock moves, the more the bouncing light pulse is forced into a stretched-out zigzag – until at the speed of light it is moving sideways as fast as it is moving up and down, and can never complete even a single bounce between the mirrors. ***Time stands still for anything moving at the speed of light.***

SPACE AND TIME COMBINED

The zigzag light clock is a particularly nice example of the working of relativity because it brings home ***the importance of looking at space and time together*** when trying to understand the implications of Einstein's special theory.

The concept of spacetime

It has become a cliché – almost as familiar as $E = mc^2$ itself – that Einstein's special theory is all about the four-dimensional geometry of SPACETIME (see page 70), a merging together of space and time. So it may come as a surprise to learn that this geometrical merging of space and time was not Einstein's idea. In fact, at first he was not impressed by it at all.

EXTENSION: (in spacetime) the four-dimensional equivalent of length

Hermann Minkowski

Hermann Minkowski (1864–1909)

Born in Lithuania (then under Russian rule), Minkowski was Professor of Mathematics at the Zurich Federal Institute of Technology when Einstein was a student there. He claimed that as a student Einstein 'never bothered about mathematics at all'. Minkowski died of appendicitis before his lecture putting the geometry into relativity appeared in print.

PICTURING RELATIVITY

* In 1905, and for years afterwards, Einstein presented his special theory in terms of algebra (equations), not in terms of geometry (pictures). The equations work perfectly well, of course - but they don't give you a physical feel for what is going on when things move at velocities that are a sizeable fraction of the speed of light.

RE-ENTER MINKOWSKI

* It was only in 1908 that Einstein's old teacher **Hermann Minkowski** – who had described Einstein as a 'lazy dog' – came up with a geometrical version of the special theory, which made it much more accessible and intelligible to non-mathematicians.

Minkowski introduced the idea of time as the 'fourth dimension', in some sense at right angles to the familiar three dimensions of space (up/down, left/right, forward/back). This led to the idea that things possess a property called EXTENSION, *which is the four-dimensional equivalent of length.*

you can make the shadow bigger or smaller than the length of the ruler

THE FOURTH DIMENSION

* A ruler, for example, has a definite length in three dimensions. But if you hold it so that it casts a shadow on the ground, you can make the shadow bigger or smaller than the length of the ruler.

* **In the same way, a ruler (or anything else) has an extension in four dimensions (in spacetime) that stays the same. But relative motion changes your perspective of the extension – as if the ruler was being twisted in spacetime.** As the ruler moves faster, *the shadow in space* gets shorter and the *shadow in time* gets longer. And when it moves more slowly, the reverse occurs (the shadow in space increases and the shadow in time decreases).

THE UNION OF SPACE AND TIME

In his lecture Minkowski made a portentous prediction: 'The views of space and time which I wish to lay before you have sprung from the soil of experimental physics, and therein lies their strength. They are radical. Henceforth space by itself, and time by itself, are doomed to fade into mere shadows, and only a kind of union of the two will preserve an independent reality.'

EINSTEIN'S GENERAL THEORY

* The general theory gets its name because it deals with motion in general, including acceleration, not just motion in a straight line at a constant speed. But it does much more than that. It explains gravity, as well. As Einstein put it, a man falling from a roof would not feel the force of gravity - he might be well aware that he was accelerating towards the ground, but he would be weightless.

Towards the general theory

From the moment when Minkowski presented his own version of the special theory of relativity, Einstein's reputation took off and in 1909 he at last left the patent office and took up his first academic post. At first, Einstein was reluctant to accept that one of his old teachers had come up with a good idea. But when, at last, he did take to the idea of the geometrization of spacetime, it helped him to develop his masterwork – the general theory of relativity.

ACCELERATION AND GRAVITY

* There are many familiar examples of what Einstein meant. The simplest is a high-speed lift. When the lift starts accelerating, as it begins to move upwards, you feel heavier and are pressed to the

come this way for a demonstration of Einstein's general theory

CONVENIENT SHORT CUT

Einstein struggled for years to find a route to a general theory, including acceleration, and later said that the happiest insight of his entire career came when he realized that gravity and acceleration are the same thing. There's no need to try to follow the mathematical complexities of Einstein's reasoning that led from this insight to a complete theory of gravity. We can jump to its finished form, using the geometrization of spacetime devised by Minkowski.

floor; when it slows down (decelerates), you feel light and rise upwards on your toes. The acceleration of the lift acts just like gravity. Not merely ***like*** gravity (we are all aware of that) but ***just like it***, because acceleration is exactly the same as gravity – which was Einstein's momentous insight.

* So far as any scientific experiments to test the theory are concerned, if you were in a steadily accelerating windowless lift, accelerating at the same rate for ever, you would not be able to tell whether you were accelerating or standing still on the surface of a planet.

Einstein realized that gravity and acceleration are the same

SPACETIME MADE INTELLIGIBLE

but I can only see three

* Most people have trouble visualizing four dimensions, so relativists use a neat trick to make spacetime more intelligible. Imagine all three dimensions of space as being represented by just one dimension. Time is now the second dimension, at right angles to this mini-space.

Unequal parts

There is an inequality between the time part of spacetime and the space part. Since the speed of light is 300,000km per second, one second of time is, in a sense, equivalent to 300,000km of space. If you had a gadget that interchanged the dimensions of space and time so you could travel in time just by walking along a road, you would have to walk 300,000km (three-quarters of the way to the Moon) in order to go back in time just one second.

WHEN MATTER IS ABSENT

* Imagine spacetime as a tautly stretched rubber sheet, very much like the surface of a trampoline. The presence of matter in spacetime is represented by a dent in the trampoline. Without any matter, spacetime is flat. If you roll a marble across the undented sheet, the marble proceeds in a straight line.

spacetime is like a taut sheet with objects resting on it

* *The mechanics of flat spacetime is, in fact, none other than the special theory of relativity. And since the general theory includes the special theory within itself, everything we have learnt about moving clocks and rulers is relevant to the general theory too.*

DENTS IN SPACETIME

* Now imagine what happens if you place a heavy bowling ball on the stretched rubber sheet to represent the Sun. With the Sun in place, spacetime is curved. If you roll a marble near the dent in spacetime made by the presence of matter, then it follows a curved path and is deflected.

a solar eclipse in 1919 gave evidence for Einstein's theory

* *Because it is spacetime itself that is bent, this deflection happens for anything moving through spacetime – any particle, obviously, but also for light. Matter tells spacetime how to bend; and spacetime tells both matter and light how to move.*

DOUBTS ECLIPSED

Einstein's general theory was published in 1916. Among other things, it suggested that any light from distant stars that happened to pass near the Sun would be deflected, very slightly, sideways by the bending of spacetime near the Sun. Usually, you cannot see stars that lie in the direction of (but far beyond) the Sun, because it is so bright. But in 1919 there was a total eclipse of the Sun, with the stars visible in daytime. Photographs taken during the course of the eclipse showed that the apparent positions of the stars had shifted slightly. The amount of the shift turned out to be exactly the amount Einstein's theory had predicted.

BOTTOMLESS PITS IN SPACE

* Continuing our trampoline analogy, imagine placing a weight on the sheet so heavy that it stretches it to breaking point, making a hole right through. If you had enough mass concentrated in a small enough region of space, you would be able to bend spacetime so much that it closed itself off from the rest of the Universe, creating a bottomless pit in the middle of flat space. This is what astronomers and physicists call a black hole.

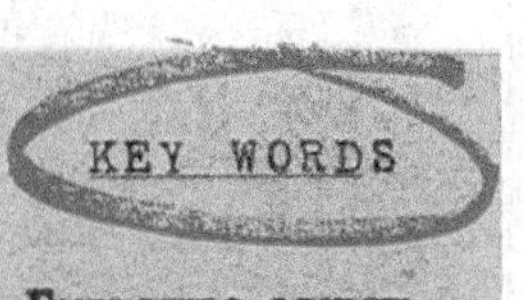

ENERGETIC OBJECT:
an object that radiates most of its energy in the form of ultraviolet light, gamma rays or X-rays
NEUTRON STAR:
tiny body with a very high density resulting from the collapse of a massive star
X-RAY STAR:
star that 'shines' at X-ray frequencies
QUASAR:
the bright central core of a galaxy (brighter than all the stars in the Milky Way put together)

ENERGETIC OBJECTS...

* The mathematics of black holes was worked out by the German astronomer **Karl Schwarzschild** (1873–1916) just before the publication of the general theory of relativity (Schwarzschild having received advance information about the theory from Einstein). Nevertheless, his work was regarded as no more than a mathematical curiosity until the 1960s, when the discovery of ENERGETIC OBJECTS in space (X-RAY STARS and QUASARS) led astronomers to think that black holes

objects around black holes radiate lots of energy and are very bright

might really exist. ***Today, they are firmly established as part of the Universe we live in.***

...AND MESSY EATERS

* High-energy objects such as X-ray stars and quasars are believed to be associated with black holes because the intense gravitational pull of a black hole attracts matter, which forms a swirling disc (called an accretion disc) around the hole. ***Material drains inward from the inner edge of this disc and is swallowed up by the hole, from which nothing – not even light – can escape.***
* But in the swirling disc, just outside the hole, particles bash together vigorously and get very hot, so they radiate X-rays, radio waves and visible light. ***Consequently, although black holes themselves are invisible, the activity surrounding them produces some of the brightest objects in the Universe.*** Black holes are messy eaters, and it shows.

Black holes

Black holes were given their name by the American physicist **John Wheeler** in 1967, but their existence had been posited decades earlier by the Indian astronomer **Subrahmanyan Chandrasekhar** (1910–95). A black hole, he suggested, would have about as much mass as our Sun packed into a ball about 3km across – which seemed nonsensical to most scientists in the 1930s. Then in the 1960s astronomers discovered neutron stars about 10km across containing as much mass as our Sun. This is so close to being a black hole that it made the idea respectable. ***Actual black holes, in the form of X-ray stars, were first identified in the 1970s.***

John Wheeler

black hole

MAKING BLACK HOLES

*** A black hole is simply a region of spacetime where gravity is strong enough to bend spacetime around, so that it's pinched off from the rest of the Universe. There are two ways you can make a black hole.**

Karl Schwarzschild

Karl Schwarzschild (1873–1916)

Schwarzschild kept in touch with scientific developments even while serving as a technical expert on the Eastern Front during the First World War. After contracting an incurable disease, he was invalided out of the army to Potsdam. There, on his death bed, he worked out the mathematics of what later became known as the theory of black holes.

METHOD NUMBER ONE

*** The first (the way Schwarzschild envisaged) is to take any mass and squeeze it into a small enough volume.*** Assuming the mass is the same, the smaller the object is then the stronger the tug of gravity is at its surface – and the more curved spacetime is. The Sun, for example, would become a black hole if it could be squeezed into a sphere just 2.9km in radius. But it is harder to do the trick for smaller masses. In order to turn

black holes swallow up matter from all around them

the Earth into a black hole, you would need to squeeze it down to just 0.88cm in radius.

METHOD NUMBER TWO

* ***The other way to make a black hole is to add more mass, thus increasing the strength of the gravitational field associated with the object without increasing its density.*** For example, imagine that you could pile stars like the Sun side by side without them merging into a single blob and shrinking under their own weight. Once you had made an object about 500 times the radius of the Sun – but with the same density as the Sun – then spacetime would once again be bent round the outside of the object so that nothing could escape. It would have become a black hole.

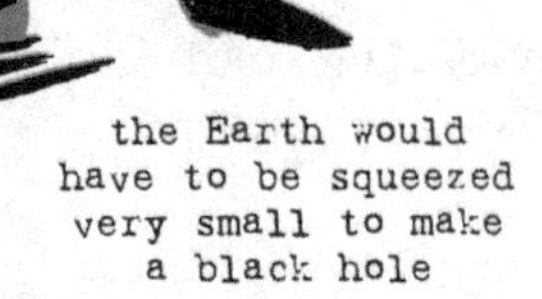

the Earth would have to be squeezed very small to make a black hole

* ***It would only have the density of water, but it would be as big across as our Solar System.*** In fact, you could make a black hole out of water, if you had enough of it – though you would need the equivalent of a few million times the mass of our Sun.

OUR LOCAL BLACK HOLE

Black holes about as big across as our entire Solar System and containing as much mass as a few million Suns are thought to exist at the hearts of some galaxies and in quasars. There may even be a slightly smaller black hole at the centre of our own Milky Way galaxy, although no longer active – probably because it has swallowed up all the matter nearby.

a large mass in a small volume would create a black hole

CAN GET IN, CAN'T GET OUT

* A key feature of a black hole is that anything can get into it, tugged by its gravitational attraction (according to the old picture) or falling down its steep gradient in spacetime (according to Einsteinian physics), but nothing can get out - not even light.

SINGULARLY DOOMED

* The equations that describe black holes describe matter plunging inexorably to its doom at a mathematical point called a SINGULARITY, at the centre of the hole. Nobody knows what happens then. At the singularity, the known laws of physics break down.

* There has been speculation that stuff falling into a singularity may get shunted through HYPERSPACE to emerge in other universes, or in another part of our own Universe. But these ideas are nothing more than speculation.

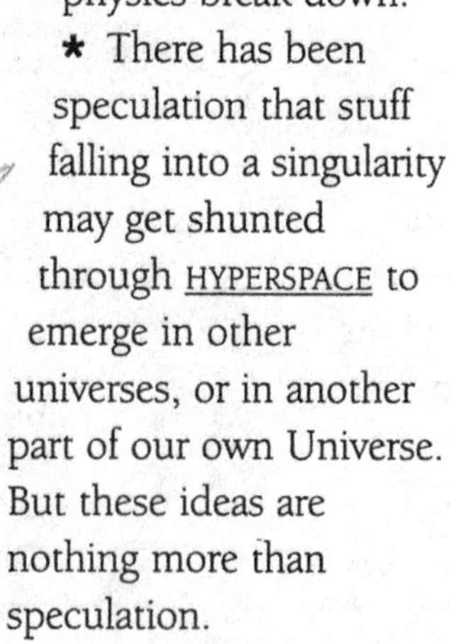

FROM BLACK HOLES TO THE UNIVERSE

At the end of the 1920s two American astronomers discovered that the galaxies beyond the Milky Way are receding from each other, which implied that the Universe is expanding. ***Einstein's general theory had, in a sense, predicted this discovery, which has remained the cornerstone of cosmology ever since. The contracting version of his equations describes what goes on in black holes. The expanding version of them describes what goes on in the Universe at large.***

EQUALLY APPLICABLE

* ***As with the other fundamental equations of physics, Einstein's equations work just as well whichever direction time is running in.*** The same equations that describe matter and radiation plunging to their doom in a singularity can be turned around to describe matter and radiation bursting out from a singularity and spreading through a burgeoning spacetime.
* ***It is these equations that describe the expanding Universe of the*** BIG BANG ***and also give what appears to be a very good description of the Universe we live in.***

the Big Bang was the outburst in which the Universe was born

KEY WORDS

SINGULARITY:
a mathematical point with zero volume

HYPERSPACE:
extension of the idea of four-dimensional spacetime to more dimensions (four-dimensional spacetime may be embedded in hyperspace, just as a two-dimensional sheet of paper is embedded in three-dimensional space)

THE BIG BANG:
the outburst in which the Universe was born – in a superhot, superdense state, possibly from a singularity

The expanding Universe

Einstein first encountered the idea of an expanding Universe in 1917. At that time, nobody knew that the Universe was expanding, so Einstein was completely baffled when his cosmological solutions to the equations of the general theory told him that spacetime could only be expanding or contracting, and could not stand still.

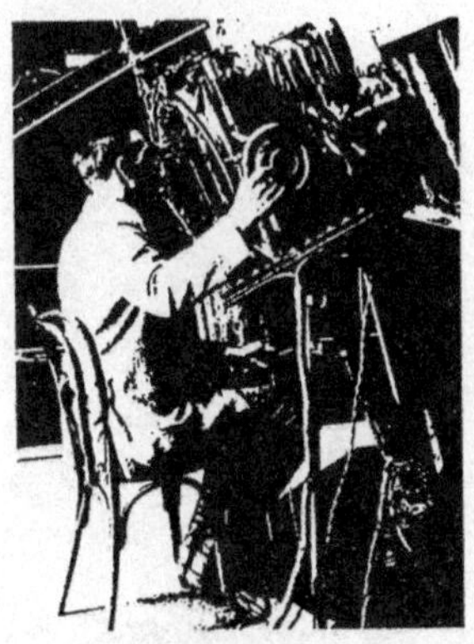

Edwin Hubble observing with the 100-inch Hooker

HUBBLE AND HUMASON

The two American astronomers who discovered that the Universe is expanding were **Edwin Hubble** (1889–1953) and **Milton Humason** (1891–1972). Hubble usually gets most credit, although Humason obtained the all-important data that Hubble interpreted. Humason was the best astronomical observer of his generation and used the best telescope on Earth at the time, the huge new reflector at Mount Wilson in California. His measurements of the spectra of light from distant galaxies revealed a RED-SHIFT EFFECT, which Hubble interpreted as evidence that all the galaxies are receding.

I want the last word on this!

BEYOND EINSTEIN

*** Even Einstein's general theory of relativity is not the last word on the subject of gravity. It can't be, because it can't describe what happens at singularities. Physicists don't know what the ultimate theory of the Universe will be. But they do have a good idea of the kind of theory that will be needed to explain such phenomena.**

Hubble and Humason established that the Universe was expanding

FRAGMENTS OF SPACE AND TIME

* The trouble with Einstein's general theory is that it is still, to a certain extent, a 'classical' theory. ***Like Newtonian mechanics or Maxwell's description of light, it deals with things that change smoothly and continuously from place to place and from time to time.***

★ Indeed, it takes no account of the other great revolution in physics that occurred during the first decades of the 20th century, the theory of QUANTUM MECHANICS – ***which describes the world of the very small (atoms and smaller) and, in a complete break from everything that went before, asserts that the world is not smooth and continuous, but is broken up into tiny pieces called*** QUANTA. And not just tiny pieces of matter, but also tiny pieces of radiation, and even tiny pieces of space and tiny pieces of time.

KEY WORDS

RED-SHIFT EFFECT: a stretching of light, caused by the expansion of the Universe, that shifts features in the visible spectrum towards the red end

QUANTUM (PLURAL QUANTA): the smallest amount of something that it is possible to have or the smallest change it is possible to make

QUANTUM MECHANICS: the rules of behaviour that apply to very small objects (atoms and smaller)

A WEIRD IDEA

★ What we need to complete our description of the world is a quantum theory of gravity – and later we will see where people are looking for such a theory. But first, what are quanta, and how did physicists come up with such a weird idea?

★ ***Once again, a revolution in scientific thinking came about because of a puzzle regarding the nature of light.*** Towards the end of the 19th century, physicists realized that if light waves really did behave like classical waves (like waves on a plucked guitar string or ripples on a pond) then they could not explain how it is radiated from a hot object, such as a candle flame, or a red-hot lump of iron, or the Sun.

another puzzle about the nature of light

you can see the visible spectrum of light in a rainbow

THE PUZZLE ABOUT LIGHT

*** A hot object radiates light (electromagnetic energy) because charged particles (electrons) are jiggling about inside it. Red light has the longest wavelength (lowest frequency) in the visible spectrum of light, which runs from red through orange, yellow, green, blue and indigo to violet. And just beyond the violet end of the visible spectrum, there is a form of electromagnetic radiation not quite visible to our eyes, called ultraviolet light.**

THE NATURE OF THE PUZZLE

The puzzle concerning the nature of light can be expressed very simply, without any mathematics. Classical physics tells you, and experiments confirm, that if you put energy into making ripples, more energy goes into making ripples with higher frequency (shorter wavelengths) than ones with lower frequency (longer wavelengths).

THE ULTRAVIOLET CATASTROPHE

* According to classical theory, preferentially any hot object ought to radiate energy at shorter wavelengths (higher frequencies) – that is, at the violet and ultraviolet end of the spectrum. ***So, when possible, any hot object would radiate all its energy away in a burst of ultraviolet radiation*** (or at even shorter wavelengths, though X-rays

blacksmith

and gamma rays were unknown at the time the puzzle was formulated). ***This became known as the ultraviolet catastrophe – a catastrophe for classical physics, because clearly nature did not behave like that.***

any hot object would radiate all its energy away in a burst of ultraviolet radiation

VISIBLY GLOWING

* What really happens when an object like a lump of iron is heated to successively higher temperatures is that at first most of the energy is radiated as infrared heat, at wavelengths just too long to be seen by our eyes. Then, it begins to glow. First it becomes red hot – then, at successively higher temperatures, orange and blue-white. The hotter the object is, the shorter the wavelengths at which most of its energy is radiated.

* The way the radiation from a hot object depends on its temperature is known as the BLACK-BODY RELATIONSHIP, because the equations that describe how energy is absorbed by a perfectly black object also describe how energy is radiated by the same object when it is heated up. ***This is another example of the way the equations of physics take no notice of the direction of time – emission of radiation is the time-reversed counterpart to absorption of radiation.***

Peak energy

The black-body relationship was worked out from experiments involving hot objects, before there was a theory to explain it. ***Puzzlingly, those experiments showed that although a little energy is radiated at longer and shorter wavelengths, the peak emission from a black body is centred on a particular narrow band of wavelength that depends only on its temperature.***

it's a trick of the light

CHAPTER 3

THE QUANTUM WORLD

★ The answer to the puzzle about the nature of light came from Max Planck, in 1900. Planck realized that the problem could be solved if the radiating objects (which we would now identify as atoms) could only emit (or, indeed, absorb) electromagnetic energy in certain fixed amounts, which he called quanta.

QUANTIFYING QUANTA

★ ***The quanta for a particular frequency of light had to carry a certain amount of energy, with the energy proportional to the frequency (and therefore inversely proportional to the wavelength) of the light involved.*** The constant of proportionality needed to make the theory fit the observations could be worked out from experiment – and is now known as Planck's constant, h. Planck stated that the energy of each quantum was given by the simple expression $E = hf$, where f is the frequency of the light being radiated.

QUANTITIES OF QUANTA

Planck thought of atoms as being able to dispense light only in certain quanta – rather like a drinks machine that is designed to dispense fruit juice in fixed quantities, to match the sizes of the cups. The machine will only give you a small cup of juice, or a medium cup, or a large one. But because you can only get drinks from the machine in three sizes, that doesn't mean fruit juice never exists in other quantities. ***It's just that the machine is set up to dispense it in these 'quanta' only.***

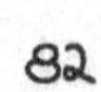

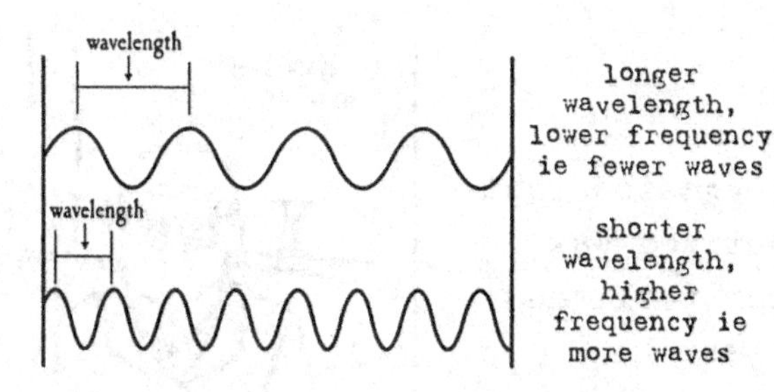

PROBLEM SOLVED

* Planck's theory solved the puzzle, because cool objects do not have enough energy to make very many high-frequency quanta. They can only radiate energy at the range of frequencies where the energy available from each atom is comparable to the energy of the quanta involved in the radiation. ***The more energy you put into the object (and the hotter it gets), the easier it is for individual atoms to radiate high-energy quanta.*** So the peak of the energy radiated (indicated by the colour of the object) shifts through the spectrum in the way previously described (see page 81).

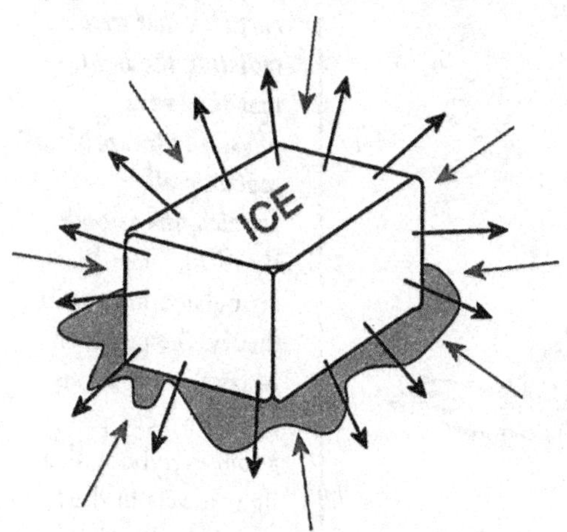

ice radiates energy, but it receives more than it gives out

Max Planck (1858–1947)

Most scientists do their best work in their twenties or thirties, but Planck was already in his forties when he made his great contribution to physics and was then still less than halfway through a full and active life. He was Professor of Physics at the University of Berlin until 1926 (when he was succeeded by Erwin Schrödinger); then in 1930, at 72, became President of the Kaiser Wilhelm Institute of Physics in Berlin. He resigned in 1937 as a protest against the Nazi regime's treatment of the Jews, but in 1945 became head of the institute again, when it moved to Göttingen and was renamed the Max Planck Institute.

Max Planck

STREAMS OF PARTICLES

* The person who suggested that light might be quantized was Einstein, in a paper published in 1905 (the same year he published the special theory of relativity). In an attempt to solve a puzzle about the behaviour of light, he harked back to Isaac Newton's idea (see page 29) that light behaves like a stream of tiny cannonballs or, in less picturesque terms, consists of particles. Although these particles were not given the name photons until the 1920s, the term can appropriately be used here.

Einstein suggested that light was a stream of particles

Planck and light

Although Planck introduced the idea of quanta into physics, he did not suggest that light *only* existed in little packets of energy. ***He thought that the quantization was a property of the particles that were radiating the light – that they were physically incapable of radiating all wavelengths at once.*** By 1900, when Planck formulated his quantum theory, the power of Maxwell's equations had convinced just about everybody that light travels in the form of a wave, like ripples on a pond.

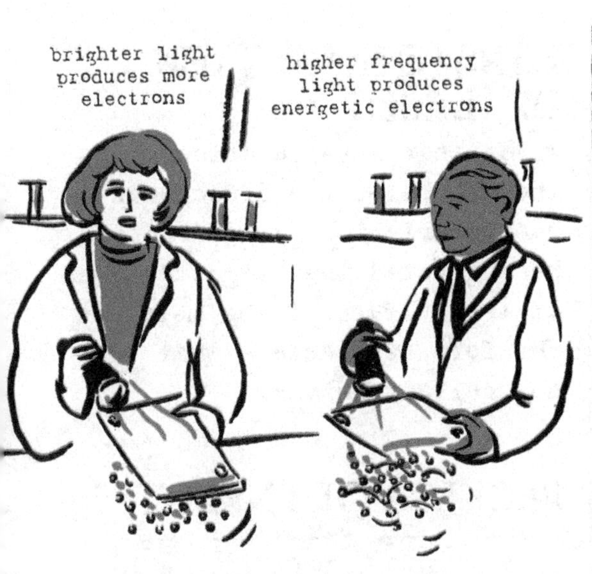

the energy of the electrons produced by shining light on metal depends on the frequency of the light not on its brightness

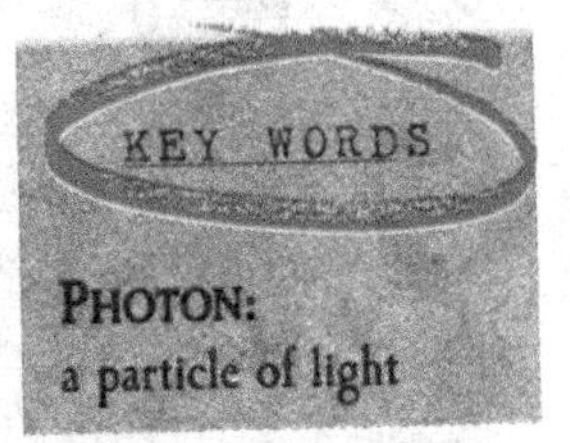

PHOTON:
a particle of light

ILLOGICAL BUT TRUE

* The puzzle Einstein solved with the aid of PHOTONS has to do with the so-called PHOTOELECTRIC EFFECT, which occurs when light shining on a metal surface knocks electrons out of the metal. The energy of the electromagnetic radiation in the light is transferred to the electrons, enabling them to escape from the grip of their parent atoms in the metal.
* *Since a bright light has more energy than a dim light, it seems logical to expect that if you shine a bright light onto the metal you will get electrons with more energy coming off. That is certainly what the pioneering experimenters expected – but that is not what happens.*

THE PHOTOELECTRIC EFFECT

The photoelectric effect is seen most clearly by using light of a particular colour. **Philipp Lenard** had been conducting experiments of this kind since 1899; and by 1905, when Einstein published his paper, they were well known. For a particular colour of light, each electron that is produced by the photoelectric effect has the same energy. For a dim light, only a few electrons are produced. For a brighter light, more are produced – but each electron has exactly the same energy as when the light is dim. ***The only way to get electrons with more energy is to use light with a higher frequency (that is, a shorter wavelength).*** Blue light stimulates the release of electrons with more energy than those released by red.

EINSTEIN'S VIEW OF LIGHT

★ Einstein's explanation of the photoelectric effect was characteristically unexpected. He suggested that light arrives at the surface of the metal in the form of quanta - what are now called photons.

Philipp Lenard (1862–1947)

As well as his work on the photoelectric effect, Lenard studied the nature of cathode rays at the end of the 1890s and, independently of J.J. Thomson, showed that they are streams of charged particles. He also came close to discovering X-rays. In the 20th century he became very bitter that all the credit for these discoveries went to Thomson and Röntgen, turned against his colleagues, and was the only leading scientist to actively support the Nazis.

Philipp Lenard

PACKETS OF ENERGY

★ Light with a particular colour is made up of photons that each have the same particular energy – little packets of energy determined by Planck's equation $E = hf$. So when each single photon gives up its energy to a single electron, each electron has the same energy. ***Bright light does carry more energy, but only because it is made up of a larger number of photons. Each of its photons still gives the same kick to the metal surface.***

★ But because blue light has a higher frequency than red light, each photon of blue light gives a bigger kick to the metal surface than each photon of red light. ***The electrons liberated by blue light therefore carry more energy than electrons liberated by red – even if you use a dim blue light and a bright red light in the experiment.***

DOUBTS AND ACCOLADES

Robert Millikan

★ ***Einstein's claim that light exists in the form of a stream of particles was so startling in 1905 that at first nobody took it seriously.*** Indeed, the American physicist **Robert Millikan** (1868–1953) spent 10 years carrying out subtle experiments designed to prove Einstein was wrong. In the end, he succeeded only in convincing himself (and everyone else) that Einstein was right – and that there was no other explanation of the photoelectric effect.

★ As a result, Einstein received the Nobel Prize (for his work on the photoelectric effect, not for either of his theories of relativity) in 1922, and Millikan got the Nobel Prize in 1923. (No need to feel sorry for Lenard – he'd already been awarded a Nobel Prize, back in 1905, for his studies of electrons.)

Einstein's strength as a scientist was his questioning attitude

EINSTEIN'S QUEST FOR TRUTH

Einstein's work on the photoelectric effect is a splendid example of how he questioned everything in his search for solutions. He had arrived at his special theory of relativity partly by rejecting Newton's view of the world and accepting the truth of Maxwell's equations – yet in the very same year, his exploration of photons and quantum theory led him to reject Maxwell's description of light and accept Newton's view of the world as correct!

THE BEHAVIOUR OF ATOMS

* By the 1920s physicists had clear proof, from the experiment with two holes, that light is a wave - and also, thanks to Lenard, Einstein and Millikan, that it consists of particles. This blurring of the distinction between waves and particles paved the way for a full theory of quantum mechanics in the 1920s. But even with only a half-baked quantum theory to work with, Niels Bohr managed to come up with a description of how individual atoms work and how they get together to form molecules.

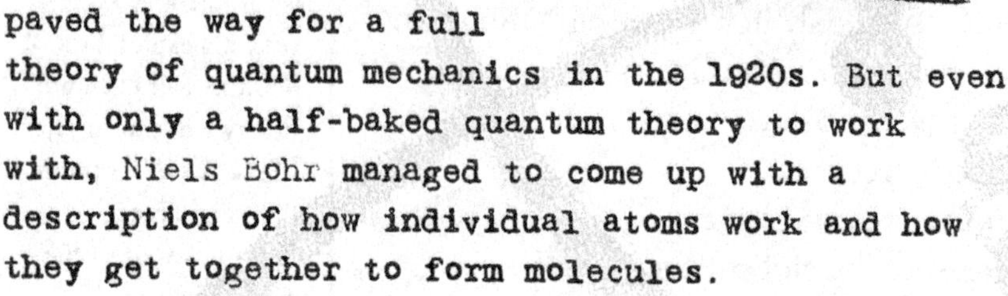

why don't electrons get attracted by the nucleus?

FOILED BY GOLD

* Bohr's description of the atom built on the experimental work of **Ernest Rutherford** and the theoretical ideas of Max Planck. In 1909 Rutherford and his colleagues at Manchester University discovered that when a beam of alpha particles was fired at a thin sheet of gold foil, although most of the particles went straight through the sheet, a few of them bounced back again.

ALPHA PARTICLES

We now know that alpha particles are the nuclei of helium atoms. Each alpha particle consists of two protons and two neutrons, held together by the strong nuclear force (see page 120). This makes a very stable unit, which behaves like a single particle in many interactions.

*** *Rutherford explained this by saying that most of the mass of an atom is concentrated in a tiny central nucleus which, like alpha particles, is positively charged – whereas the electrons associated with the atom, which carry negative charge (so each atom is electrically neutral overall), are somehow distributed in a cloud around the nucleus.***

* Consequently when an alpha particle hits the electron cloud, it brushes through it without noticing. But if it happens to head straight for the nucleus, then the electric repulsion of the two sets of positive charge pushes it back the way it came or off at a sharp angle.

WHY NO ATTRACTION?

* From the statistics of the experiment, Rutherford was able to work out that the nucleus is 100,000 times smaller than the electron cloud around an atom. But this left a big puzzle. ***Why didn't the electrons (which have negative charge) get attracted by the nucleus (which has positive charge) and fall into it?***

Three kinds of radiation

At the beginning of the 20th century, Ernest Rutherford gave the names 'alpha rays' and 'beta rays' to the two kinds of radiation produced by radioactivity. A little later, when a third kind of radiation was discovered, it was called 'gamma rays'. Beta rays turned out to be fast-moving electrons. Gamma rays are electromagnetic radiation, like X-rays but even more energetic. Alpha rays are also particles (see opposite).

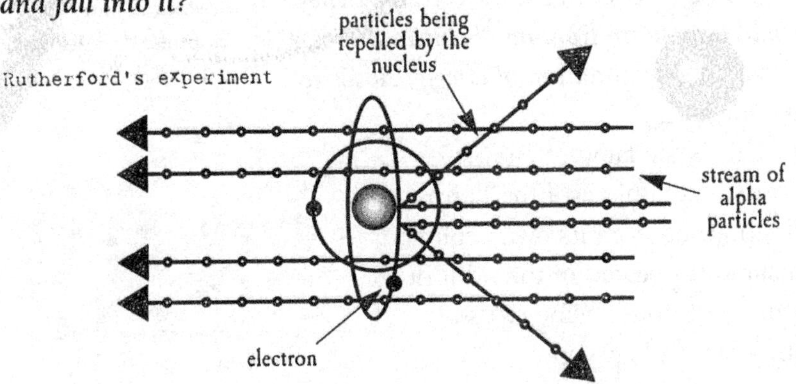

Rutherford's experiment

at the centre of each atom there is a nucleus of neutrons and protons

ELECTRONS IN ORBIT

* Bohr's idea was that the electrons must, in some sense, be 'in orbit' around the nucleus, rather like the planets orbiting around the Sun. However, this wasn't enough to stabilize the atom - because an accelerated electric charge radiates energy, and circular motion constitutes an acceleration. According to classical physics (including Maxwell's equations), the electrons orbiting in an atom would radiate energy away and spiral into the nucleus.

Niels Bohr (1885–1962)

After completing his PhD, Bohr worked with Rutherford's group at Manchester University, where he came up with his model of the atom. By 1918, he was so highly regarded that he was invited to head a new institute of physics in Copenhagen (now called the Niels Bohr Institute). Most of the great theoretical physicists of the day visited his institute at various times, and in the 1920s it was a catalyst for the development of the theory of quantum mechanics – which is how the Copenhagen Interpretation got its name.

THE QUANTUM LEAP

* Max Planck had shown that electromagnetic radiation could only be radiated in quanta. Bohr now suggested that the electrons in the atom could only radiate whole quanta of light, not smaller bits. So they couldn't spiral inwards. ***They could only jump from one orbit to the next – exactly one quantum of energy closer to the nucleus.***

* This is the famous 'QUANTUM LEAP'. It is rather as if the Earth disappeared from its own orbit and instantly appeared in the orbit of Venus, without having crossed the space in between.

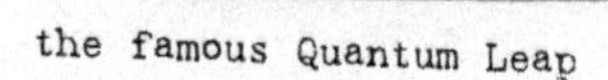

the famous Quantum Leap

WHY THE LEAP?

* But why don't the electrons simply jump down step by step, as if bouncing down a staircase, until they all pile up on the bottom step?

* *Bohr's answer was to say that there must be a rule of quantum physics that only allows a certain number of electrons in each orbit. There is only room for two electrons in the orbit closest to the nucleus. An atom of lithium, for example, which has three electrons would therefore have to put the third one in the next orbit out from the nucleus, a quantum of energy further away.*

Rutherford and his laboratory

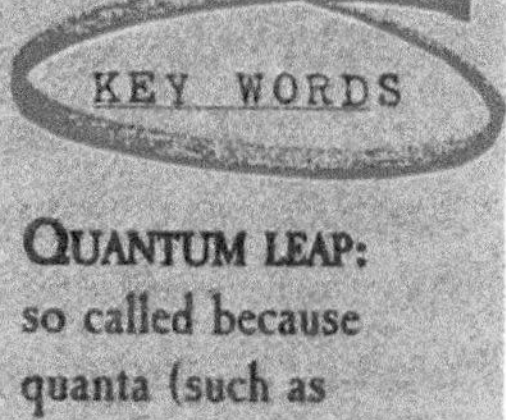

QUANTUM LEAP: so called because quanta (such as atoms) 'leap' abruptly from one state to another

Ernest Rutherford (1871–1937)

Born in New Zealand, Rutherford worked in England, then Canada, then back in England, where he eventually became the head of the Cavendish Laboratory. Together with **Frederick Soddy**, he showed that radioactivity involves atoms of one element being transformed into atoms of another. In 1909 he devised the experiment, carried out by **Hans Geiger** and **Ernest Marsden**, that led to the discovery of the atomic nucleus. When Rutherford received the Nobel Prize, in 1908, it was for chemistry (for his work on elements), although he considered chemistry an inferior discipline and once said that 'all of science is either physics or stamp collecting'.

BOHR'S MODEL OF THE ATOM

* The reason why Bohr's model was greeted as a significant step forward was that by spacing the electron orbits out in the way demanded by Planck's equation, he could explain the spectrum of light from a simple atom.

the electron orbits are like a flight of stairs

KEY WORDS

BOND:
a link between two atoms that involves sharing or exchanging electrons

SHELL:
a zone (within an atom) in which a certain number of electrons are allowed to be present

PROTON:
positively charged particle found in the nucleus of an atom

NEUTRON:
similar to a proton, but having zero electric charge

Niels Bohr

JUMPING UP AND DOWN

* ***All atoms radiate or absorb light at specific wavelengths, making sharply defined lines in the spectrum.*** The pattern made by these lines (sometimes known as the spectral signature) is unique for each element – as unique as a fingerprint.

* ***Bohr showed that each line corresponds to an electron jumping from one orbit (or one 'energy level') to another – or rather, many identical electrons making the***

identical jump in many atoms of the same element. A jump down releases energy and produces a bright line, while if the atom absorbs energy it makes a dark line in the spectrum as the electron jumps upwards.

* The important point is that the jumps always involve emission or absorption of a whole number of quanta – an amount of energy equivalent to hf, or some multiple of hf, but never any amount in between.

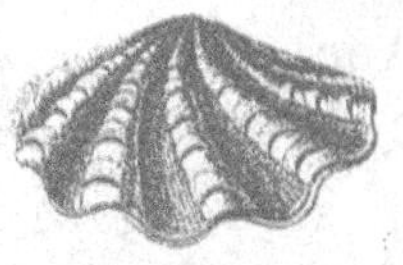

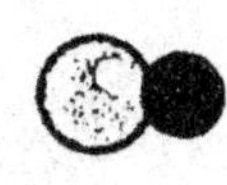

the molecules themselves are made up of atoms

BONDS AND SHELLS

* By the early 1920s Bohr had developed his model of the structure of the atom to explain the basics of chemistry – how and why atoms join together to make molecules.

* ***A link that holds atoms together to make molecules is called a*** BOND. ***The electrons in different orbits are said to occupy*** 'SHELLS' ***– with different energy levels – around the central nucleus of an atom. The shells are like layered onion skins, each containing a certain number of electrons.***

oxygen molecule O_2

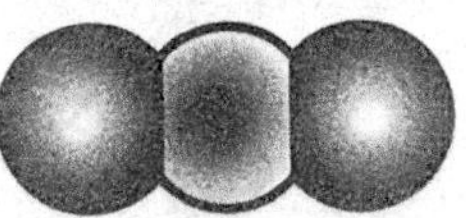

carbon dioxide molecule CO_2

FILLING THE SHELLS

The most stable arrangement for each shell occurs when it contains the maximum number of electrons allowed by the quantum rules, which is two for the innermost shell and eight for most others. The simplest atom, hydrogen, has one PROTON in its nucleus and one electron in orbit around it. The next element, helium, has two protons and also a couple of NEUTRONS, orbited by two electrons in the same shell. But, as we have seen, lithium has to put its third electron in the next shell, and so on.

one carbon and four hydrogen should do it

ATOMIC BONDS

*** Bonding results from the atoms' attempts to reach the desirable state of having a full outer shell. One way the optimum distribution of electrons between shells can be achieved is by sharing electrons between two atoms.**

we could agree to share...

Electronic illusion

Water is another example of a covalent bond. Two hydrogen atoms, each possessing a single electron, link up with one oxygen atom, which has six electrons in its outermost occupied shell. Again, each atom has the illusion of a filled shell.

COVALENT BOND

* Each hydrogen atom has one electron, and would 'like' to have two to fill its only occupied shell. Each carbon atom has six electrons, two in its (full) innermost shell and four in its (half-empty) outer shell. If four hydrogen atoms surround a carbon atom in the right way, the four hydrogen atoms each get a part-share in one of the outer four electrons of the carbon atom; and the carbon atom gets a part-share in each of the four electrons associated with the hydrogen atoms. Consequently, all of the atoms in the resulting molecule – CH_4 (methane) – have the illusion of a full outer shell. ***This is called a covalent bond.***

hydrogen atom

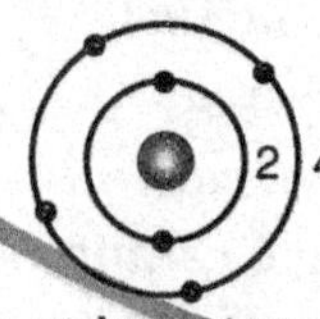

carbon atom

IONIC BOND

★ Another type of bond keeps sodium and chlorine bound together in common salt. Each sodium atom has 11 electrons: two in a full innermost shell, eight in another full shell, and one on its own outside. If it could lose the outermost electron, it would be left with a full shell as its visible face. Chlorine atoms, on the other hand, each have 17 electrons: two in a full innermost shell, eight in the next (full) shell, and seven in the outermost occupied shell. They need one additional electron to fill the outermost shell.

the sodium atom loses one electron, the chlorine atom gains one electron

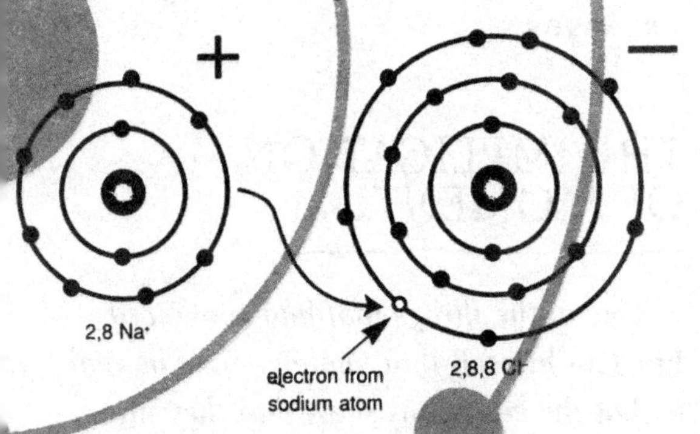

★ When sodium and chlorine combine, each sodium atom in effect gives up one electron to a chlorine atom, so both have obtained a full outermost occupied shell. But this leaves each sodium atom with one unit of positive charge; and each chlorine atom with one unit of negative charge. ***The charged particles are held together by electric forces, arranged in a*** <u>CRYSTAL LATTICE</u>. ***This is called an*** <u>IONIC BOND</u>.

KEY WORDS

COVALENT BOND: the kind of bond formed when pairs of electrons are shared between two atoms

IONIC BOND: the kind of bond formed when electrons are transferred from one atom to another

CRYSTAL LATTICE: an array of atoms held in place in a rigid geometrical structure by electric forces

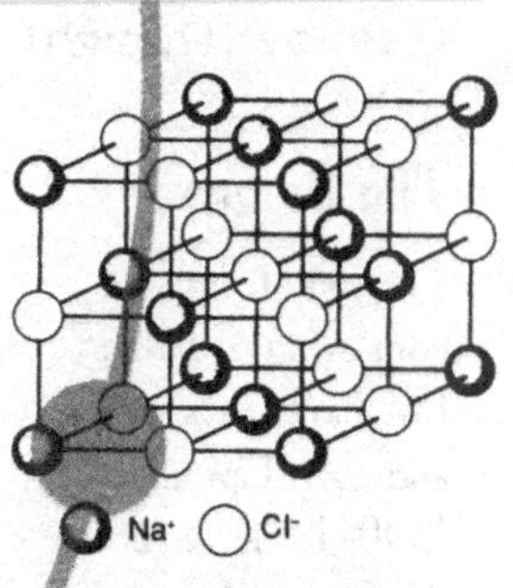

sodium chloride lattice

DE BROGLIE'S INSIGHT

* Bohr had used a mixture of classical and quantum ideas, plus ad hoc rules about the behaviour of electrons based on spectroscopy and other experiments. It was a mish-mash, but worked after a fashion. Then in the mid-1920s Louis de Broglie pointed out that if things like light that had been thought of as waves also behaved like particles (photons), maybe things regarded as particles (electrons and the like) should also be thought of as waves.

Louis de Broglie

The Braggs

The only team consisting of father (William, 1862–1942) and son (Lawrence, 1890–1971) to receive the Nobel Prize for physics for work they carried out together. The work involved the diffraction of X-rays by crystal lattices, which revealed details of the structure of crystals. When the prize was awarded to them, in 1915, Lawrence was only 25 and was serving in the army in France.

THE IMPLICATION OF MOMENTUM

* *One of the things that had convinced Einstein himself that photons must be real is that the equations show that they must carry momentum, just like other particles.*

* The momentum of a particle of light is a prediction of the special theory of relativity, and is found by dividing its energy by its speed. Planck had shown that the energy of a light quantum is equal to hf. And the speed of light is just c. So the momentum of a photon is hf/c. Momentum is usually denoted by the letter p (because m is used for mass) and

wavelength by the Greek letter l (lambda). The frequency of a wave divided by its speed is equal to one over its wavelength, so we can write $p = h/l$ or $pl = h$.

* This was true for light. But electrons also have momentum – ***so de Broglie realized that they must obey the same equation***, and that an electron with momentum p must have a wavelength given by the equation $l = h/p$.

A MOMENTOUS IMPLICATION

* De Broglie's hypothesis was soon confirmed by experiments in which beams of electrons were fired at crystal lattices – just as X-rays had been fired at crystal lattices by **William and Lawrence Bragg**, the pioneers of X-ray crystallography.
* Like the X-rays, or like light in the experiment with two holes, the electrons interfered with one another and produced a DIFFRACTION PATTERN, proving that they were waves. ***All quantum entities, it was realized, not just light, have this dual wave-particle nature.***

What did it mean?

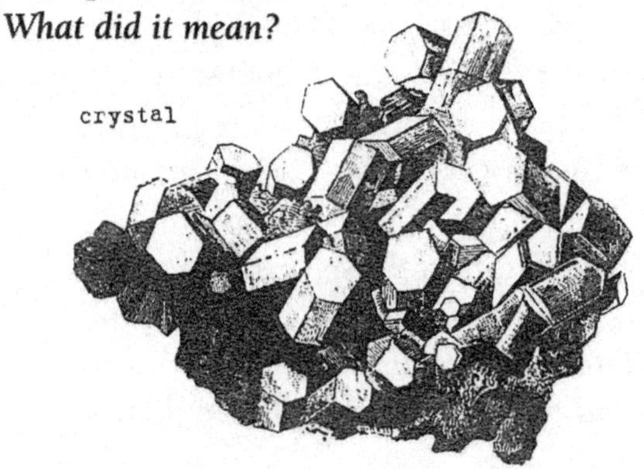

Louis de Broglie (1892–1987)

De Broglie was a French aristocrat, whose family intended for him to join the diplomatic corps. But, encouraged by his older brother Maurice, he turned to physics. His studies were interrupted by army service during the First World War, when he worked on radio communications and was for a time based at the Eiffel Tower. As a result, he did not complete his PhD, presenting his wave theory of electrons, until 1924, when he was already in his thirties. Louis and Maurice both served on the French High Commission of Atomic Energy, and were involved in the peaceful applications of atomic energy after the Second World War.

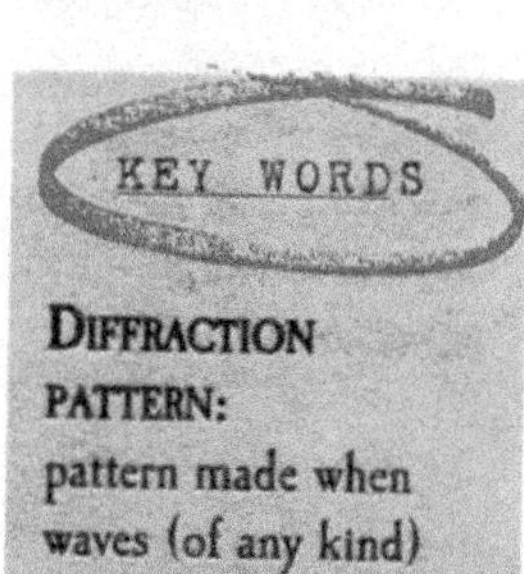
KEY WORDS

DIFFRACTION PATTERN: pattern made when waves (of any kind) bend round an obstruction or spread out from a small hole

THE EXPERIMENT WITH TWO HOLES REVISITED

* The experiment with two holes has now been refined to such an extent that it is possible to fire individual photons one at a time, so each of them arrives at the detector screen on the other side and makes a pinpoint of light. It is also possible to do the same kind of thing with electrons, using a detector screen like a TV screen where each electron makes a single point of light as it arrives.

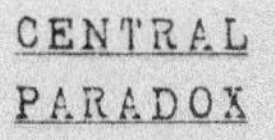

The experiment with two holes incorporates what Richard Feynman (see page 145) used to call 'the central mystery' of quantum mechanics: ***all particles are waves, and all waves are particles.***

SPONTANEOUS INTERFERENCE...

* Both versions of the experiment have actually been done. In both cases, the single quantum entities (photons and electrons) start out on one side of the experiment as particles and arrive at the other side as particles.

* *But if you run the experiment many times, either with photons or with electrons, then the spots of light on the detector screen add up to produce an interference pattern. The implication is that both the photons and the electrons*

pass through the two holes as waves, and somehow interfere with themselves to make the interference pattern.

...AND TELEPATHIC PARTICLES?

* Stranger still, each individual particle travels through the experiment on its own. And makes the spot of light on the screen on its own. ***But if you send hundreds or thousands of particles through the experiment, not even all together but one after the other, they conspire to produce an interference pattern.*** Each quantum entity seems to be aware of the whole experimental set-up – including the presence of the two holes in the middle of the experiment, and the presence of the preceding and following particles.

* ***Quantum entities somehow seem to take no notice of space and time – or rather to take account of all of space and all of time at once.***

Revolutionary physics

The development of the theory of quantum physics in the 1920s has been described by Nobel Laureate Steven Weinberg as 'the most profound revolution in physical theory since the birth of modern physics in the 17th century'. It 'changed our idea of the questions we are allowed to ask'.

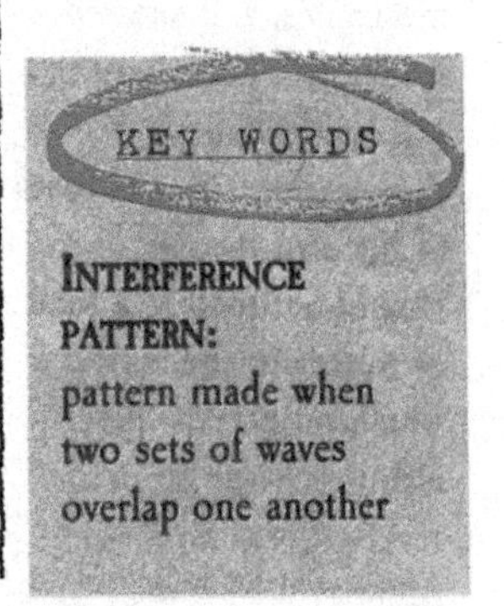

KEY WORDS

INTERFERENCE PATTERN: pattern made when two sets of waves overlap one another

Werner Heisenberg

I don't agree with waves

Erwin Schrödinger

RIVAL THEORIES

*** In 1925 two different teams of physicists came up with two different ways to describe all this mathematically. The one approach, which was pioneered by** Werner Heisenberg, **discarded the idea of waves and offered a solution based on quantum jumping. The other, pioneered by** Erwin Schrödinger, **deliberately tried to get rid of what he called 'this damned quantum jumping' and used the idea of waves.**

Paul Dirac (1902–84)

Probably the greatest English scientist since Newton, Dirac was so self-effacing that his name is almost unknown outside scientific circles. He came up with the most complete early version of quantum mechanics, and showed that both Heisenberg's and Schrödinger's theories were special cases of his own theory. The idea of antimatter, a kind of mirror-image matter in which properties like electric charge are reversed, was also originated by him.

THAT DAMNED JUMPING!

* ***Strangely, both approaches to the quantum world gave exactly the same answers when used to calculate things like the behaviour of electrons in atoms – and both, of course, agreed with the findings of the experimenters.***
* Within a year, **Paul Dirac** had shown why. Both of the approaches were mathematically equivalent to one another – ***because each of them was a special case of a more abstract kind of mathematical formalism that didn't involve the***

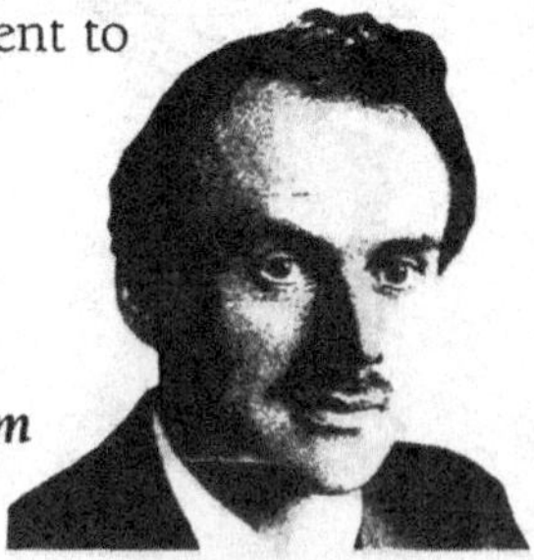

Paul Dirac

image of quantum entities as either waves or particles (or as anything we are at all familiar with).

* This discovery – with its implication that even his wave equation could not get rid of that 'damned quantum jumping' – led Schrödinger to express his disgust with the theory he had helped to develop, saying, 'I don't like it, and I'm sorry I ever had anything to do with it.'

QUANTUM REALITY?

* But Niels Bohr, Max Born and others seized on the new mathematical equations of quantum mechanics and produced a picture of what seemed to be going on in the quantum world. The picture didn't make sense – at least not in terms of everyday common sense – but it did have its own internal logic and it did seem to describe quantum reality.

the Allies were concerned that Heisenberg might make an atomic bomb for the Nazis

Werner Heisenberg

FEVERISH EFFORT

Werner Heisenberg (1901–76) was able to work out his version of quantum mechanics, in the spring of 1925, thanks to a vicious attack of hay fever. Recuperating on the island of Heligoland, without distractions, he had the opportunity to develop his radically new vision of the way the world worked. Since Heisenberg was the only prominent physicist of his generation to stay at his post in Germany during the Second World War, the Allies feared he might be leading a Nazi programme to develop an atomic bomb. However, he always maintained that he steered the Nazis away from atomic weapons. The truth will never be known.

HEISENBERG'S UNCERTAINTY PRINCIPLE

One of the most famous aspects of the quantum world is Heisenberg's uncertainty principle. Heisenberg used the wave-particle duality to show that you can never measure both the exact position of an entity such as an electron and its exact momentum at the same time. ***In other words, you cannot simultaneously know where the electron is and where it is going. Worse still, the electron itself can't know both where it is and where it is going (at least not exactly).*** Quantum uncertainty isn't anything to do with the imperfection of our experiments or the deficiencies of our measuring equipment – it is built into the very fabric of the quantum world.

WHAT IS MEANT BY UNCERTAINTY

* To understand what Heisenberg meant by uncertainty, think about the properties of waves (also, remember that all quantum entities have a wave nature). A wave cannot possibly be located at a single point in space, because by its very nature a wave is a spread-out thing. A pure wave (with a single wavelength) stretches out for ever and carries momentum, but has no location and no resemblance to a particle. In contrast, a particle has a definite position but it has no wave properties.

WAVE PACKETS

* *However, if you mix different waves together, you can make what is called a* WAVE PACKET, *where the waves cancel each other out everywhere except in a small region*. The wave packet is a bit like a particle – but with a rather fuzzy boundary and no exact location. It can be made as small as you like, up to a point (or rather, not quite literally up to a point).

wave packet

a wave has momentum but it has no definite location, because it is spread out

Momentum and position

Heisenberg showed that there is a very precise trade-off between momentum and position. If you multiply the amount by which the position of an electron (or anything else) is uncertain by the amount by which its momentum is uncertain, the number you get is always bigger than Planck's constant, *h*.

* Now think about momentum. A single pure wave has a definite momentum, which can be determined by the equation Einstein worked out from his special theory of relativity. ***But in a wave packet there is a mixture of waves and a mixture of momenta. In order to make the wave packet smaller, you need more waves in the packet, to do the crucial cancelling out, and that creates more uncertainty about the momentum of the wave packet.***

Heisenberg said you cannot simultaneously know where the electron is and where it is going

THE STRANGENESS OF THE QUANTUM WORLD

* Now you can see why the strange features of the quantum world don't show up in everyday life, and are not common sense. But the reason why we don't have any doubt about the position of a bag of sugar on a shelf, or the momentum of a snooker ball rolling across a table, is that they are so big - compared with things like electrons.

it's all so strange

Talking about quanta

When we talk about a quantum entity being ***like*** a wave or ***like*** a particle, we mean just that – and nothing more. All we are doing is using our everyday experience, and everyday language, to provide us with an image of what is going on.

TINY DIFFERENCES

* Planck's constant (*h*) is absolutely tiny. Using the system of units where energy is measured in ergs and mass in grams, $h = 6.6 \times 10^{-27}$ (that is, a decimal point followed by 26 zeros before the 66). Quantum effects only become important for entities that have masses (measured in grams) in this sort of range, or smaller. The mass of an electron, for example, is 9×10^{-28} grams.

* An uncertainty of *h* in the momentum or position of an electron is a big deal – whereas a

$h = 6.6 \times 10^{27}$ (that is, a decimal point followed by 26 zeros before the 66)

similar uncertainty in the position or momentum of a bag of sugar or a snooker ball, or a human being, is far too small to be noticed at all.

waves are an everyday way of thinking about quantum physics, but in reality quantum particles are not like anything we can see

MARCHING TO THE QUANTUM RULES

★ ***Uncertainty and wave-particle duality are intimately linked – but the important thing to remember about all the strangeness of the quantum world is that nobody is saying that a thing like an electron 'really is' a particle, or that it 'really is' a wave.*** Its very nature is different from anything we are used to in everyday life. It obeys a different set of rules – the quantum rules – and marches to the beat of a different drum.

PICTURING UNCERTAINTY

Most people are unhappy with the idea of uncertainty, because they can't get a picture of it in their heads. They are happier with the idea of a wave, because we have all seen waves. This is why the wave version of quantum mechanics is the version most people find easiest to use. ***But electrons (and other quantum entities) are not waves – they merely behave like waves, in some circumstances.***

QUANTUM THEORY IN ACTION

*** A nice example of the way in which quantum entities differ from everyday entities, and how the wave picture helps as an analogy, is a phenomenon known as the HYDROGEN BOND.**

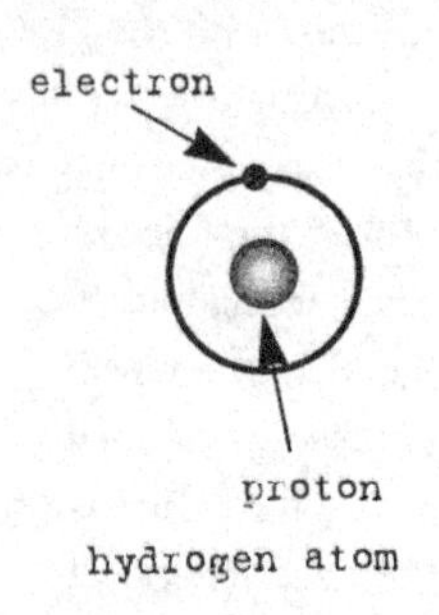

hydrogen atom

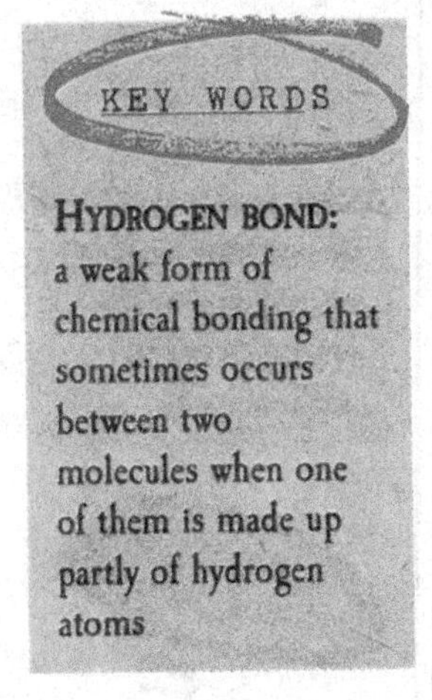

HYDROGEN BOND: a weak form of chemical bonding that sometimes occurs between two molecules when one of them is made up partly of hydrogen atoms

ALTERNATIVE PERCEPTIONS

* A hydrogen atom consists of a single proton (the nucleus), surrounded by a single electron. The words 'surrounded by' are completely appropriate. One way of looking at things is to say that the position of the electron is uncertain – it could be anywhere in a cloud surrounding the proton. Alternatively, instead of talking about a tiny particle in orbit around the proton, you could say that there is a wave that winds right round the atom. Either way, the hydrogen atom behaves as if the electric charge carried by the electron is spread out uniformly around the proton.

The ratio of hydrogen to oxygen in water. There are two atoms of hydrogen for every atom of oxygen

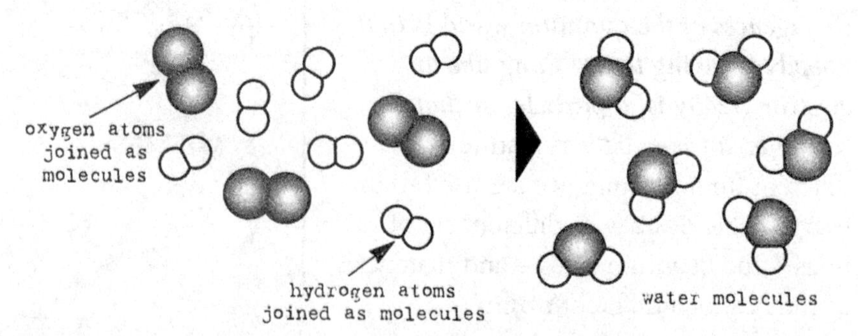

THE NICETIES OF THE HYDROGEN BOND

* When a hydrogen atom links up with something else to form a molecule – for example, when two hydrogen atoms join onto a single atom of oxygen to make water (H_2O) – the electron cloud is gripped both by the hydrogen atom's nucleus and by the nucleus of the other atom. It forms a bridge between the two atoms. ***But because the electron is now concentrated at one end of the hydrogen atom, the other end is partly uncovered.***

* If electrons really were little particles, and chemical bonds formed in exactly the way Bohr described, the other end of the hydrogen atom would be completely bare, exposing its proton's positive charge. So water molecules would interact strongly with almost anything they came into contact with. But because even a single electron is spread out around the hydrogen nucleus, there is still a little shielding of the positive charge on the hydrogen nucleus. ***This means that the end of the hydrogen atom opposite the chemical bond acts as if it has a fraction of a positive electric charge.***

* Although the quantum theory of hydrogen bonds was worked out by **Linus Pauling** (1901–94) in the 1930s, it was only in the late 1990s that experiments became subtle enough to measure all the details of the effect he predicted and confirm that he was right.

PREDICTABLE BEHAVIOUR

This difference between how particles behave and how quantum entities behave produces a measurable effect, in terms of the strength with which the partly exposed proton interacts with other atoms and molecules. ***The strength of this interaction can be measured in chemical experiments – and matches the predictions of quantum theory, not classical ideas.***

THE COPENHAGEN INTERPRETATION

* The 'explanation' of the quantum world that Bohr and others developed is called the Copenhagen Interpretation, because a lot of it was worked out at Niels Bohr's institute in that city. We have already encountered two of the main planks of the Copenhagen Interpretation, uncertainty and wave-particle duality. Another key idea, the role of PROBABILITY in quantum processes, was introduced by Max Born.

Max Born

you are looking lovely tonight

Complementarity

Wave-particle duality is sometimes known as COMPLEMENTARITY, because the two aspects of quantum reality complement each other. You can't have one without the other.

THOSE TWO HOLES AGAIN

* In a quantum system, you can never be sure of the outcome of an experiment. If there are several possible outcomes, ***all you can do is work out the probability of each one*** – exactly like working out the odds of a particular hand being dealt at cards. In the simplest example, when an electron confronts the two holes in the experiment with two holes, ***provided nobody looks***, it seems to go both ways at once and interfere with itself. ***But if you set up a detector at each of the holes to monitor the electrons***

working out the odds

passing by, you find that half of them go one way and half go the other way – with no interference.

if you look you will spoil the experiment

DISTURBING OBSERVATION

* ***This is also an example of the fourth plank of the Copenhagen Interpretation – which asserts that the system is disturbed by the very act of observing or measuring it.***

* This idea can be seen at work in the COLLAPSE OF THE WAVE FUNCTION. An electron wave function is quite happy going through both holes at once if nobody looks at it. ***But if you look, it collapses into a particle and goes one way or the other – with, in this particular case, equal probability.*** There's a 50:50 chance that any individual electron will be seen going through one hole or the other one, but you can never predict from the laws of physics which hole it will choose.

ONE OR OTHER

* The collapse of the wave function doesn't matter if you are designing something practical like the picture tube of a TV set, because billions of electrons are involved. ***Provided half go one way and half go the other way, you don't care what each individual electron does.***

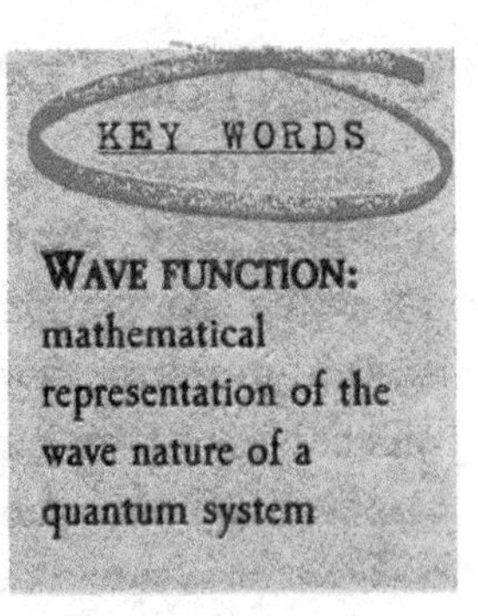

KEY WORDS

WAVE FUNCTION: mathematical representation of the wave nature of a quantum system

Max Born (1882–1970)

Max Born, who introduced the concept of quantum probability, was Professor of Physics at Göttingen University from 1921 to 1933. A warm and friendly man, he made Göttingen a leading centre of quantum physics and was greatly irritated when the standard theory (to which he made major contributions) became known as the Copenhagen Interpretation. Forced to leave Germany when Hitler came to power, during the 1930s he lectured at Cambridge then became Professor of Natural Philosophy at Edinburgh University.

QUANTUM PROBABILITY

*** This business of quantum probability shows up very clearly in** RADIOACTIVE DECAY **- which occurs when an unstable atom (strictly speaking, an unstable nucleus) spits out particles and transforms itself into something else.**

'I cannot believe that God plays dice.'

ALBERT EINSTEIN

A DICEY BUSINESS

* ***Each kind of*** RADIOACTIVE ATOM ***has a characteristic lifetime, known as its*** HALF-LIFE. ***In any pile of radioactive atoms of the same kind, exactly half will transform themselves in this way in one half-life.*** Half the rest (a quarter of the original) will decay in the next half-life, and so on. Yet any individual atom in the pile might decay at once, or it could stick around for very many half-lives before decaying – exactly as if each atom tosses a coin, repeatedly, and decides to decay only if the coin comes up heads.

* It was the role of probability in quantum mechanics, and in the Copenhagen Interpretation in particular, that led Einstein to make the famous remark, 'I cannot believe that God plays dice.' ***But all the evidence suggests that Einstein was wrong: nature (what Einstein, an atheist, always meant by 'God') operates on the same principles as a casino.***

THE VALUE OF QUANTUM PHYSICS

***It is important to appreciate that the package of ideas enshrined in the Copenhagen Interpretation really does work.** It is the toolkit most practical physicists use when they are dealing with quantum entities – which, for practical physicists, usually means electrons and/or light.

* Far from being some abstract byway of science, quantum physics underpins all of the modern understanding of chemistry. To take the most profound example, the famous double-helix structure of DNA – the molecule of life – is held together by hydrogen bonds, which are a quintessentially quantum phenomenon.

* More obviously, quantum physics also describes the behaviour of electrons moving through circuits and semiconductors, including the microchips at the heart of modern computers. And it describes the behaviour of lasers, those intense pulses of light now used prosaically in everyday items such as CD players.

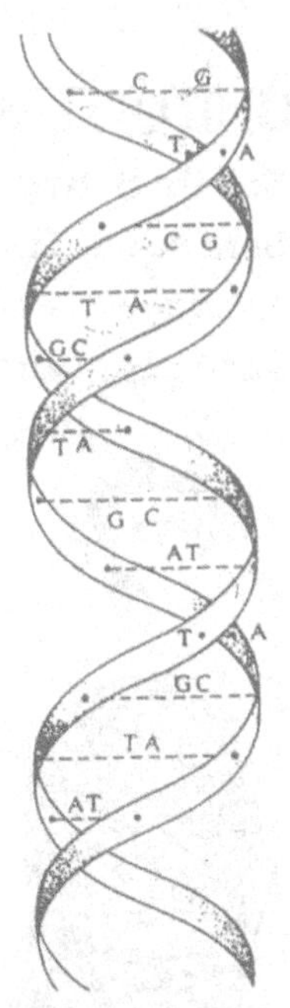

DNA double helix

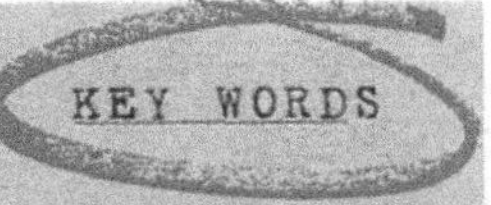

RADIOACTIVE ATOM: an unstable atom which may spit out a particle (usually an electron or an alpha particle), turning itself into an atom of another element

RADIOACTIVE DECAY: the process whereby a radioactive atom spits out a particle

HALF-LIFE: the time it takes for half the radioactive atoms in a sample to decay

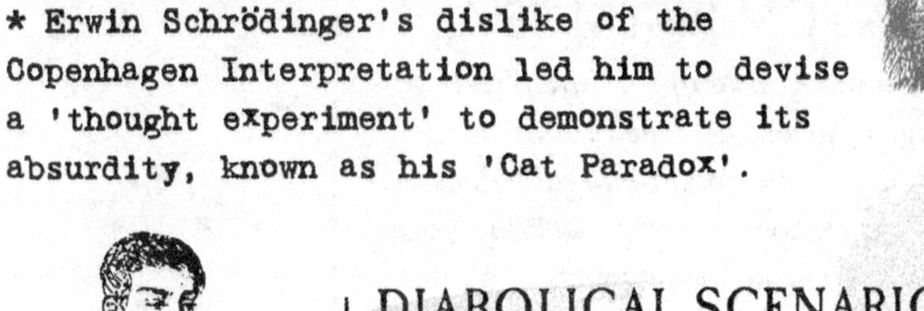

SCHRÖDINGER'S CAT

★ Erwin Schrödinger's dislike of the Copenhagen Interpretation led him to devise a 'thought experiment' to demonstrate its absurdity, known as his 'Cat Paradox'.

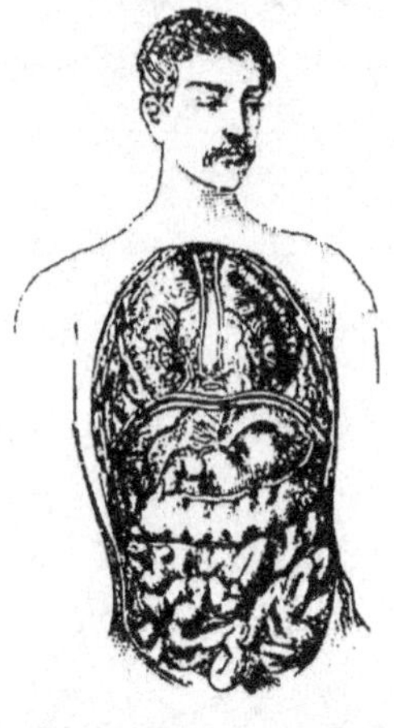

INDIGESTIBLE PHYSICS

Quantum physics is weird, but it works. Which doesn't make the Copenhagen Interpretation any easier to digest. The person who found it most difficult to stomach was Erwin Schrödinger. That was why he came up with his 'Cat Paradox' in 1935. But his experiment is not so much a paradox as a graphic illustration of the weirdness of the quantum world.

DIABOLICAL SCENARIO

★ Schrödinger asks us to imagine a sealed chamber in which there is a cat, supplied with food and drink and all the other things it needs for a comfortable and healthy life. But in the chamber there is also a 'diabolical device', which is hooked up to a sample of radioactive material (perhaps just a single atom). ***The ingenious device is set up so that if and when the radioactive material decays, it will trigger the release of poison gas, which will kill the cat.***

will the cat be alive?

CRUCIAL QUESTIONS

* If you wait outside the chamber for a suitable length of time, there will come a moment when there is exactly a 50:50 chance that the radioactive sample has decayed in just the right way to trigger the diabolical device. ***What, asks Schrödinger, is the state of the cat at that precise instant? And what will you see if you open the chamber door?***
* The second question is easier to answer. If you look, you will see either a dead cat or a live cat, with equal probability. Or to put it another way, if you were to do such an experiment a thousand times, half the time you would find a dead cat and half the time you would find a live one.

Schrödinger's cat experiment shows the weirdness of quantum physics

Erwin Schrödinger

Erwin Schrödinger (1887-1961)

The man who came up with the most widely used version of quantum theory, wave mechanics, thought he was restoring sanity to a subject that had got out of control. A physicist of the old school, he was nearly 40 when he did this work. By expressing quantum physics in terms of waves, Schrödinger planned to bring it in line with classical ideas. When this proved impossible, he was deeply upset – and had little more to do with quantum theory, except to point out what he believed to be its flaws.

I'm not going to look..

I'm not going to look.

OPENING THE DOOR

*** According to the Copenhagen Interpretation, the collapse of the wave function *only occurs when you look* - so *just before* you open the door of Schrödinger's hypothetical chamber and look inside, the radioactive sample is still hovering in a state of uncertainty, not sure whether it has decayed or not.**

DISCLAIMER

It is usual to comfort the reader by stressing that no experiment like the one in Schrödinger's puzzle has ever been carried out and no cats have been mistreated in this way – the idea of a thought experiment being that it is literally all in the mind.

DEAD OR ALIVE?

* At this point its wave function still contains both possibilities, like the unobserved electron in the two-hole experiment going through both holes at once. ***So the poison both has and has not been released, and the cat is both dead and alive.*** Everything inside the chamber is supposed to be described by an uncertain wave function, which only collapses when somebody looks at it. Nobody believes the inside of the chamber is really like that – but the trouble is that nobody has come up with a better way, mathematically, to describe what is going on.

LETTING THE CAT DECIDE

* One seemingly obvious way to resolve Schrödinger's paradox is to say that the cat is quite capable of deciding whether or not it is alive, and can collapse the wave function all by itself. But then where do you draw the line? Can you do the same sort of thing with a rat, or a flea, or a microbe?

* ***Some people just say that it is the number of atoms (or other particles) involved that makes more complicated systems behave in accordance with classical physics, while simpler systems (that is, ones having fewer atoms) obey quantum rules.***

well I'm not going to do it

* But again, where do you draw the line? DNA molecules obey quantum rules. Fleas don't. Where in between is the boundary? ***Can there really be a system that obeys quantum rules but which if you added just one more atom to it would start to behave classically? Nobody can say.***

DNA molecules WILL obey quantum rules

What is life?

Schrödinger left Germany in 1933, when the Nazis came to power, and turned up in Oxford with both a wife and a mistress, but he didn't stay long. After several moves, he settled in Dublin, where an institute was set up to give him a base. There he wrote a book about the molecular basis of life, entitled *What is Life?* This encouraged a generation of physicists to turn to molecular biology after the Second World War. One of them was Francis Crick, co-discoverer of the structure of DNA.

MANY WORLDS

* To some people, Schrödinger's puzzle is so worrying that they have come up with all kinds of alternative 'interpretations' of the quantum world. But in their own ways they are all just as weird as the Copenhagen Interpretation; for instance the 'Many Worlds' Interpretation, which will be familiar to science-fiction fans.

First, but not best?

Some people find the 'Many Worlds' idea easier to accept than the Copenhagen Interpretation. Some don't. But more people use the Copenhagen Interpretation than any other interpretation, simply because it came first and was established in the marketplace by Niels Bohr's powerful advocacy of it.

SEPARATE REALITIES

* According to the 'Many Worlds' Interpretation, every time the quantum world is faced with a choice (such as whether or not the radioactive sample in the chamber with the cat decays), the entire Universe splits into two – or into however many copies are required to cover every possible outcome of the quantum event.

* In the case of Schrödinger's cat, instead of hovering in a state of uncertainty, the radioactive atom both decays and doesn't decay, creating two separate realities – which in some sense coexist alongside each other.

* ***These separate realities are complete copies of each other in every detail except for the outcome of the particular quantum choice.*** So there is one Universe in which you

open the door and find a dead cat, and a Universe next door in which an exact copy of you opens the door and finds a live cat.

MULTIPLE INTERPRETATIONS

computer chip

* The thing is, all of the different interpretations of quantum mechanics give you exactly the same answers about things you can actually measure. They have exactly the same value in designing computer chips, or tinkering with DNA for the purposes of genetic engineering, or whatever. If there was any practical way to tell them apart at this level, it would be clear which one was best – ***but because they are all mathematically identical, you are allowed to choose which one you like (or which you find least unacceptable).***

the 'Many Worlds' theory says there are parallel worlds, one where the cat is dead and one where it is alive

MATHEMATICAL TRUTH

Don't worry if you can't understand how the quantum world works. In the words of Richard Feynman (who won the Nobel Prize, in 1965, for his work on quantum physics and knew what he was talking about), 'nobody knows how it can be like that'. ***What matters is the truth about the way the world works, expressed in mathematics.*** The truth lies in the equations – and all the physical images are just mental crutches, designed to help inadequate human minds get a picture of what is going on by using analogies drawn from the world of everyday experience. And that, surely, is all anyone without a degree in physics needs to know about quantum mechanics.

CHAPTER 4

EVEN NEWER PHYSICS

* The latest physics forms a two-pronged attack on the puzzle of how the world works. One line of attack continues the approach developed by the quantum pioneers, probing deeper and deeper into the structure of matter in an attempt to find the ultimate building blocks. The other takes a different route entirely, and concentrates on the way large numbers of building blocks (whatever they may be) behave. The suggestion is that the whole may be greater than the sum of its parts - or at least, that it would be a good idea to find out whether or not this is true.

FIELD QUANTA

The old physics said that particles are carried from one particle to another by waves in a field of force. The new physics says they are carried by other particles, called field quanta.

FERMIONS AND BOSONS

* The everyday world is made of atoms, and the behaviour of the entities that make up atoms is very well described by the rules of quantum physics. Although we should remember that all particles are also waves, and all waves are also particles, it is still convenient to think of the material world as made of particles and to talk about the interactions between particles as being associated with fields like the electromagnetic field.

★ But there is a fundamental difference between the two ways of looking at things. The things we are used to thinking of as particles, such as electrons, are conserved (there is always the same number of electrons in the Universe). Whereas the things we are used to thinking of as field quanta – such as photons – are not conserved (you make billions of them when you turn on a light, but they disappear when they are absorbed by your eye). For historical reasons, the first kind of particles are called FERMIONS, *while the latter (the photon-like particles) are called* BOSONS.

there is always the same number of electrons in the Universe

QUANTUM ELECTROMAGNETISM

★ The particles that make up atoms are the protons and neutrons in the nucleus and the electrons surrounding the nucleus. The interactions between the electrically charged particles (the protons and electrons) and between atoms themselves are described by the quantum version of electromagnetism – a theory known as QUANTUM ELECTRODYNAMICS, or QED.

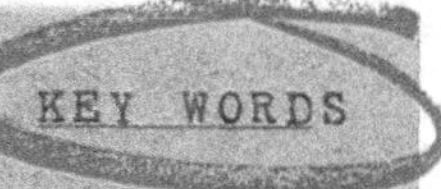

FERMIONS: particles of the material world – things like electrons and protons

BOSONS: force-carrying field quanta – things like photons

All change

Instead of describing electromagnetic interactions in terms of waves, in QED these interactions are described in terms of photons being exchanged between charged particles. **In quantum theory, because all waves can be described as particles, it follows that all forces can be described in this way (as the exchange of field quanta). The photon is the quantum of the electromagnetic field.**

I'll exchange you a couple of photons for that

a force of strength keeping them together

GRAVITY AND GRAVITONS

We don't have to worry about gravity when thinking about the forces between atoms and molecules at the level of chemistry, because gravity is very much weaker than electromagnetism. To be precise, the electrical force of repulsion between two protons is about 10^{36} (a 1 followed by 36 zeros) times stronger than the gravitational force trying to pull them together. That is why stars are so big. They have to be, in order for gravity to overwhelm those electromagnetic forces and make protons fuse together in their cores, releasing energy as they do so. ***But gravity can also be described as operating by the exchange of field quanta – which in this case are called*** GRAVITONS.

NUCLEAR INTERACTIONS

*** In order for nuclei of atoms to hold together despite the fact that they are filled with positively charged protons, there has to be a force much stronger than electromagnetism - and much, much stronger than gravity - to do the trick. It is called the strong nuclear force.**

proton

electrons moving around the nucleus

nucleus

neutron

inside the atom

THE STRONG NUCLEAR FORCE

* ***The strong nuclear force has a very short range. In fact, it cannot be felt at a distance bigger than the diameter of a nucleus – which is why nuclei have the diameters they do.*** But it is indeed very much stronger than the electromagnetic force (about a hundred times stronger) and holds the nucleus together in spite of the repulsion of all that positive charge. This situation is also helped by the presence in the nucleus of neutrons, which also feel the strong force, but have no electric charge.

THE WEAK NUCLEAR INTERACTION

* There is one other interaction that affects nuclei. This is called the weak nuclear interaction – because it is weaker than the strong nuclear interaction and weaker than electromagnetism (although it is still 10^{25} times stronger than gravity). ***It operates in a subtly different way from the other interactions and makes itself known by causing the process of radioactive decay, when an unstable nucleus splits (or fissions) into two or more pieces.***

BETA DECAY

The weak interaction is also the driving force behind a related process known as beta decay. This happens when a neutron left on its own spontaneously spits out an electron (electrons used to be known as beta rays), plus a particle called an electron ANTINEUTRINO, and transforms itself into a proton. That is allowed by the quantum rules – even though electrons are conserved – because the electron antineutrino balances the books. For the purposes of this kind of book-keeping, the neutron and proton can be regarded as different versions of the same particle, which is sometimes dubbed the NUCLEON.

KEY WORDS

GRAVITON: the field quantum of the gravitational field

NEUTRINO: a particle in the same family as the electron that has no electric charge and only a tiny mass (possibly no mass at all)

ANTINEUTRINO: antimatter counterpart to the neutrino

NUCLEON: either of the two kinds of particle found in an atomic nucleus – the proton and the neutron

atomic book-keeping is all about keeping the positive and negative charge balanced

PARTICLES AND ANTIPARTICLES

* ANTIPARTICLES are a kind of mirror-image counterpart to everyday particles: they possess reversed properties. For example, the antiparticle counterpart to the electron, the anti-electron, has positive charge instead of negative charge. For this reason, it is often called the POSITRON.

the antiparticles are a mirror image

KEY WORDS

ANTIPARTICLE: a 'mirror-image' version of a particle that has reversed properties

POSITRON: the antiparticle corresponding to an electron

QUARK: a fundamental particle that exists inside protons, neutrons and some other particles

THE DISCOVERY OF QUARKS

* There is no experimental evidence that electrons have any internal structure (they behave as if their mass and electric charge were concentrated at a point). But in the 1960s several lines of experimental evidence indirectly suggested that there might be structure inside protons and neutrons. **George Zweig** ***and*** **Murray Gell-Mann** ***independently came up with an explanation for***

we call it a quark!

the experiments by proposing that there were particles inside both kinds of nucleons. Gell-Mann gave these particles the name 'QUARKS'. Although he intended the name to rhyme with 'pork', most physicists pronounce it to rhyme with 'bark' – but either pronunciation is acceptable.

Murray Gell-Mann

INSIDE NUCLEONS

* In the 1970s a team at the Stanford Linear Accelerator, in California, probed the structure of nucleons by firing beams of electrons at solid targets – in much the same way that Rutherford's team had probed the structure of the atom by firing alpha particles at gold foil in 1909 (see page 88). The electrons bounced off something inside the nucleons, and James Bjorken and Richard Feynman showed that the properties of the particles inside nucleons matched the expected properties of quarks.

* ***The picture that emerged of the structure inside nucleons is that the proton and the neutron are each composed of three quarks. But only two different kinds of quark are needed to describe them.*** These quarks were given the whimsical names 'UP' and 'DOWN' – but the names don't mean anything, they could just as well have been called Tom and Jerry. ***A proton is made up of two up quarks and one down quark, while a neutron is made up of two down quarks and one up quark.***

COUNTING PARTICLES

When we are working out how many particles are involved in an interaction, each particle counts as +1 and each antiparticle counts as –1. ***There must always be the same number of particles coming out at the end of the interaction as went in at the beginning.*** In this sense, when a neutron decays into a proton and releases an electron and an antineutrino, these two entities cancel each other out. We start with a single particle, the neutron, and a count of +1. We end up with three 'particles', but the count is +1 for the proton, +1 for the electron, and –1 for the antineutrino. So the total is still +1.

COLOURFUL NAMES

The name given to the property of quarks analogous to electric charge is totally arbitrary. It doesn't mean that quarks are coloured in the everyday sense of the word, any more than calling a child Rose means she is rose-coloured. It is just a name. It might have been better if the colour charge and colour field had been called the glue charge and glue field, especially since the term 'glue force' is used. Although there are only two types of electric charge (positive and negative), there are three kinds of colour charge: RED, GREEN and BLUE. They could just as well have been called Huey, Dewey and Louie.

THE COLOUR FORCE

* The quarks are held together inside nucleons by what seems at first sight to be another force (or interaction), which behaves in a similar way to the strong nuclear force. But it turned out that this is really the underlying interaction responsible for the strong nuclear force itself - so there are still only four fundamental forces of nature.

COLOUR CHARGE

* Quarks carry an electric charge (the up quark has a charge of $^2/_3$ of the charge on a proton, and the down quark has a charge of $^1/_3$ of the charge on an electron), but

they also have another property very similar to electric charge. It is called colour charge, but the name is entirely arbitrary. It doesn't mean that quarks are actually coloured.

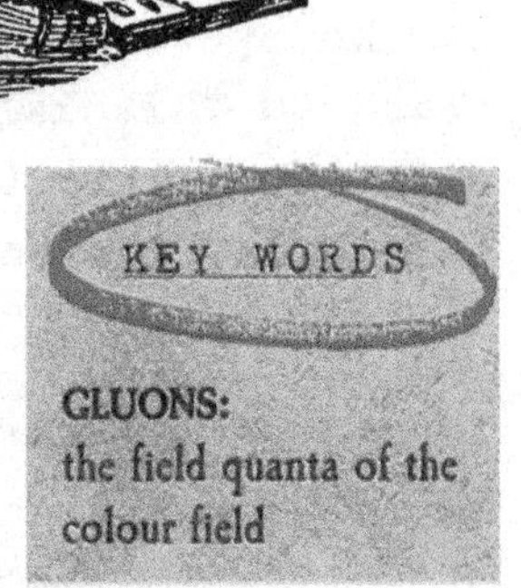

QUANTUM CHROMODYNAMICS

★ ***Quarks are held together by the exchange of particles called*** GLUONS ***(because they 'glue' quarks together), which play the role in the context of the*** COLOUR FIELD ***that photons play for the electromagnetic field.***

★ The quark/gluon theory that describes this behaviour is modelled on QED (see page 119) and is called QUANTUM CHROMODYNAMICS, or QCD. Some of the colour force that holds quarks together inside nucleons leaks out of the nucleon, and can hold on to the nucleon next door. ***And this is what we perceive as the strong nuclear force.***

quantum physics is full of whimsical names

KEY WORDS

GLUONS:
the field quanta of the colour field

Strange duplication

The two kinds of quarks needed to make protons and neutrons are neatly balanced by the electron and its partner the neutrino, which is involved in processes like beta decay (see page 121). ***There are just four kinds of particle needed to describe all of the observed material world, and just four kinds of force. Strangely, though, it turns out that the particles are duplicated in nature – not once, but twice.***

ACCELERATED PARTICLES

* Particle accelerators are machines used to speed particles like electrons and protons to very high energies and smash them together. We talk about high energies, rather than high speeds, because no particle can move faster than the speed of light. As the speed of the particles in accelerators gets close to this limit, the energy put in (from electromagnetic fields) increases the mass of the particles, instead of making them go faster.

new particles are created in the collisions

Charm and strange, top and bottom

As well as the up and down quarks, there are heavier, unstable, versions that can be manufactured in particle collisions – namely, a pair known as CHARM and STRANGE, and a pair known as TOP and BOTTOM. Nobody knows why nature should have triplicated the fundamental particles in this way, making three 'generations' of the fundamental particles. But, for most purposes, all we need to worry about are the up and down quarks, the electron and its neutrino, and the four fundamental interactions.

CREATING NEW PARTICLES

* New particles are created in the collisions. ***These are literally new, not just pieces of colliding particles that have broken off.*** The new particles are made out of pure energy – in line with Einstein's famous equation $E = mc^2$, which can be turned around to read $m = E/c^2$. The more energy (E) you put in, the more mass (m) you can get out (as ever, c is the speed of light). ***Most of the particles made in this way are unstable, and soon decay into familiar forms such as electrons and protons.***

three types of electrons

HEAVY VERSIONS

***** *As well as the electron and its neutrino, it is possible by using a particle accelerator to manufacture a heavier version of each particle, and a third, even heavier, version.*

***** A heavy electron is called a MUON. Its mass is just over 200 times bigger than the mass of an electron (for comparison, the mass of a proton is about 2,000 times the mass of an electron, and the neutron has much the same mass as a proton), but it is otherwise identical to an electron. Although it can be made in accelerators, it decays into an electron in just two-millionths of a second.

SPLIT-SECOND LIVING

The even heavier electron is called a TAU PARTICLE (sometimes tau-minus), or TAUON. It has a mass 70 per cent bigger than a proton, but its lifetime is only 30-thousandths of a billionth of a second. The muon and the tauon each have their own kind of neutrino to accompany them in particle interactions (neutrinos have only tiny masses). All these electron-like particles and neutrinos are collectively known as LEPTONS.

KEY WORDS

MUON:
a heavier version of the electron

TAUON:
an even heavier version of the electron

LEPTONS:
collective name for the electron, muon, tauon and their respective neutrinos

THE WEINBERG-SALAM THEORY

*** One of the greatest achievements of theoretical physics in recent decades came in 1967, when two researchers,** Abdus Salam **and** Steven Weinberg, **each independently found a way to combine the electromagnetic interaction and the weak nuclear interaction in one package. They showed that both interactions could be regarded as manifestations of a single force, which became known as the ELECTROWEAK INTERACTION.**

physicists are on a quest for a universal theory of everything

POPULAR PHYSICIST

Outside the field of theoretical physics Steven Weinberg, the American physicist who played a major role in the development of the electroweak theory in the 1960s, is known to a wide audience as the author of a best-selling book about the Big Bang, entitled *The First Three Minutes.*

ELECTROWEAK AT WORK

* ***The crucial feature of the electroweak theory is that it includes four field quanta.*** There is the familiar photon, which mediates the electromagnetic interaction; and there are three other bosons, called W^+, W^- and Z^0, which between them mediate all possible weak interactions.

* There are three of these bosons because weak interactions, unlike electromagnetic interactions, can involve the particles concerned in changing their electric charge

– as when a neutron (which is electrically neutral) decays into a positively charged proton and a negatively charged electron, plus an electron antineutrino. According to the new picture, what happens is that the neutron spits out a W^- particle, leaving a proton behind, and the W^- then decays into the other two particles.

* More generally, the charged quanta of the electroweak field can carry either negative electric charge or positive electric charge from one particle to another. But since some weak interactions do not involve a change of charge, there also has to be a neutral boson, the Z particle.

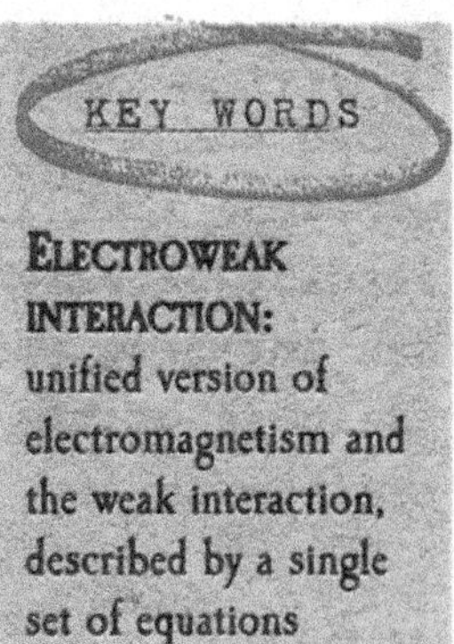

PREDICTIONS FULFILLED

* Unlike photons, though, the two Ws and the Z particle have to have mass. ***The great triumph of the electroweak theory was that when these particles were first manufactured in accelerator experiments, at CERN in the early 1980s, the masses predicted for the particles turned out to be exactly correct.*** The two Ws each have about 83 times the mass of a proton, while the mass of the Z is about 93 times a proton's. All three particles are, of course, highly unstable and quickly decay into other things.

it's about 83 or could it be 93 times the mass

Quest for a universal theory

What physicists would like to have is a single theory that can describe all of these entities in one set of mathematical equations, in the same way that Maxwell's equations described everything known to 19th-century science about electromagnetic interactions. This is a good analogy, because Maxwell showed that electricity and magnetism could be unified in one mathematical package. If two forces can be unified in this way, why not all of them?

MADE FROM PURE ENERGY

★ How can a particle like a neutron spit out a particle that is 80 or 90 times heavier than itself? Because the particle that is being spat out has been made out of pure energy, not out of the mass of the neutron. It was never 'inside' the neutron in any sense.

Abdus Salam (1926–96)

Devoted to encouraging the development of science in the Third World, Salam worked in Pakistan, where he was born, then in Cambridge, before becoming Professor of Theoretical Physics at Imperial College, London (1957–93). Besides his pioneering work on electroweak theory, he set up the International Centre for Theoretical Physics, in Trieste, where young scientists from developing countries work alongside visiting experts.

Abdus Salam

ENERGY FROM UNCERTAINTY

very strong forces have a short range, but less strong forces have an enormous range

★ *The energy comes from* QUANTUM UNCERTAINTY. *Like the uncertainty that relates position and momentum, there is an uncertainty that relates time and energy. It says that for a brief instant of time, a packet of energy can pop into existence out of nothing at all.* The shorter the time involved, the more energy is allowed to appear in this way. But it has to disappear again within a time limit determined by its equivalent mass – which is given by that familiar equation $E = mc^2$.

PROMPT DISCARD

* Enough energy to make a W or a Z particle can appear in this way, provided it promptly disappears. In beta decay, a little extra energy comes from the mass of the neutron ($E = mc^2$ again) – so when the W^- particle is forced to disappear, it turns the extra energy it stole from the neutron into an electron and its antineutrino partner, instead of leaving behind nothing at all. Protons are therefore a bit less massive than neutrons.

* ***The amount of energy involved determines how far a particle made in this way (called a*** VIRTUAL PARTICLE***) can travel before it has to decay. The more mass, the shorter the distance it can go. This limits the range of the bosons involved and explains why the strong and weak nuclear forces only operate on a tiny scale, whereas the electromagnetic interaction and gravity – which are mediated by massless photons and gravitons respectively – have an enormous (in principle, infinite) range.***

a packet of energy can appear from nothing at all, provided it promptly disappears

can I borrow a bit of energy until Tuesday

KEY WORDS

QUANTUM UNCERTAINTY: Phenomenon of the quantum world that applies to various pairs of properties, such as position-momentum and energy-time. The more accurately one member of the pair is known, the less accurately the other is known.

VIRTUAL PARTICLE: A particle that only exists as a result of quantum uncertainty. It 'borrows' energy from the vacuum to make itself, but quickly has to give it back and disappear. Empty space is a seething froth of virtual particles.

ELASTIC BANDS AND GLUE

* Unlike gravity and electromagnetic forces, the glue force is *strongest* when it operates on pairs or triplets of quarks that are far apart. This sounds utterly bizarre, until you realize that we are all familiar with something that acts in a similar way - an elastic band. If the two ends of the band are close together, the elastic is loose and floppy; but if you move the ends apart, the elastic stretches and resists the movement by pulling back.

KEY WORDS

THE GLUE FORCE: another name for the colour force that holds quarks together in pairs and triplets
CONFINEMENT: term describing the fact that quarks are always confined inside other particles
ANTIQUARK: antimatter counterpart to a quark
PION: particle made up of a quark-antiquark pair

HOW THE GLUE FORCE WORKS

* The glue force is like that. If you try to move two quarks apart (perhaps by hitting one of the quarks inside a proton with a high-speed electron), they can only go so far before being snapped back to where they belong – in this case inside the proton. This is why the glue force only has a short range in practice, even though gluons have no mass.

A GREGARIOUS EXISTENCE

Because of a property called CONFINEMENT, quarks are never seen in isolation. This happens because of the way the GLUE FORCE operates. The quantum rules only allow pairs or triplets of quarks to exist.

QUARK PLUS ANTIQUARK EQUALS PION

* If you pull an elastic band hard enough, it will break. With a pair or triplet of quarks, if you hit one hard enough it will eventually break free, in a sense.
* But this can only happen if you put so much energy in that there is enough to manufacture not one but *two* new quarks (strictly speaking, a quark and an ANTIQUARK) at the point where the break occurs. One of the new quarks snaps back into place inside the proton, and the other new quark (actually the antiquark) escapes alongside the first quark, in a quark-antiquark pair known as a PION. ***No experiment has ever shown the existence of an isolated quark, but when matter is bombarded by high energy electrons it is possible to see streams of pions emerging.***

Three interactions or four?

Strictly speaking, even today we really only have three fundamental interactions (the electroweak interaction, QED and gravity) to worry about, not four. However, many people have tried to take things a stage further by bringing QCD into the fold. There are several different ideas about how this might be done, and they are generally known as GRAND UNIFIED THEORIES, or GUTs.

I've got a GUT feeling about this

GRAND UNIFIED THEORIES

*** Nobody has quite succeeded in finding a grand unified theory that works perfectly - largely because QCD is a much more complicated kind of interaction than QED or the weak interaction.**

QED AND QCD

* In QED (see page 119) there is only one kind of boson to worry about – the photon – and it doesn't have any mass. In the weak interaction there are three bosons to worry about, and they have mass as well. But at least none of these bosons interact with each other.

* In QCD (see page 125) there are eight different kinds of bosons – the gluons – to worry about. They don't have mass (which is one less thing to worry about) but they can interact with each other, even forming entities known as GLUEBALLS. Which makes things a lot more complicated.

EQUALITY FOR LEPTONS

*** *If a grand unified theory could be developed, it would have to include quarks and leptons on an equal footing.*** This is similar to the way the electroweak theory puts electrons and neutrinos on an equal footing, with the appropriate bosons able, thanks to their electric charge and other

properties, to change electrons into neutrinos, and vice versa. (This is essentially what happens in the second stage of beta decay, after the neutron has spat out a W^- particle.)

FAMILY LIKENESS

* The equality of quarks and leptons seems likely, because, as we have seen, there are three generations of quarks and three generations of leptons, hinting at some sort of family connection.

* ***So there would have to be a kind of boson that could transform quarks into leptons, in the same way that the weak interaction can transform a neutrino into an electron – or turn itself into an electron plus an antineutrino, which is the same thing.*** The various approaches that have been developed towards a grand unified theory predict the mass of this field quantum, which is called the X-BOSON.

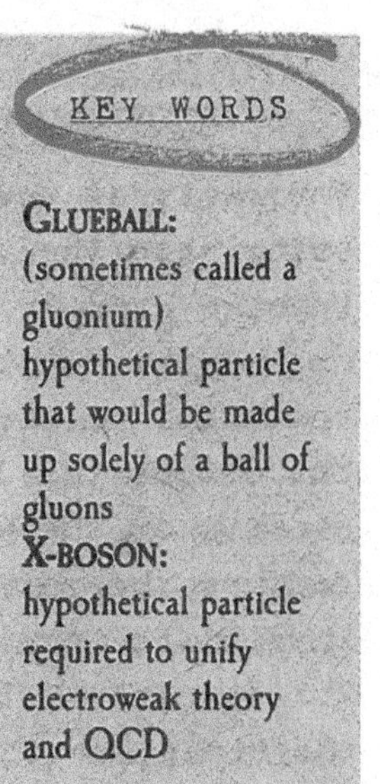
KEY WORDS

GLUEBALL: (sometimes called a gluonium) hypothetical particle that would be made up solely of a ball of gluons

X-BOSON: hypothetical particle required to unify electroweak theory and QCD

A difficult test

Since different GUTs predict different masses for the X-boson, why don't we test the theories by measuring the mass of real X-bosons? Unfortunately, all the masses predicted by the candidate GUTs are so high (around a million billion times the mass of a proton) that there is no hope of manufacturing them in any particle accelerator that could be built on Earth. In fact the energies required to make such massive particles would only have existed in the first split second of the existence of the Universe – in the Big Bang.

TOWARDS A THEORY OF EVERYTHING

* Gravity is even harder to bring into the fold, because gravity is so much weaker than all the other interactions. But during the last two decades of the 20th century theoretical physicists found a way to circumvent all the difficulties with QCD and gravity, and jump straight to a complete theory of all of the particles and forces of nature - a true 'THEORY OF EVERYTHING' (or TOE), operating at a much deeper level than anything we have considered so far.

Cancelling out the infinities

In theories like QED, the infinities are got rid of by a trick called RENORMALIZATION. This amounts to dividing one infinity by another one to make them cancel out. It works, but the cleverest physicists, such as Paul Dirac and Richard Feynman, were always deeply unhappy about it.

ZERO SIZE

* The prehistory of this new idea – which goes by the prosaic name of STRING THEORY – takes us back to the 1960s. ***Not surprisingly, up to the end of the 1960s particle physicists had pictured fundamental particles like electrons simply as particles.***

* Some (such as protons, which have a measurable size) were thought of as tiny objects like miniature billiard balls.

billard balls

Others, such as electrons (which have no size measured by any experiment to date), were thought of as points with literally zero size. But there were always problems with this picture.

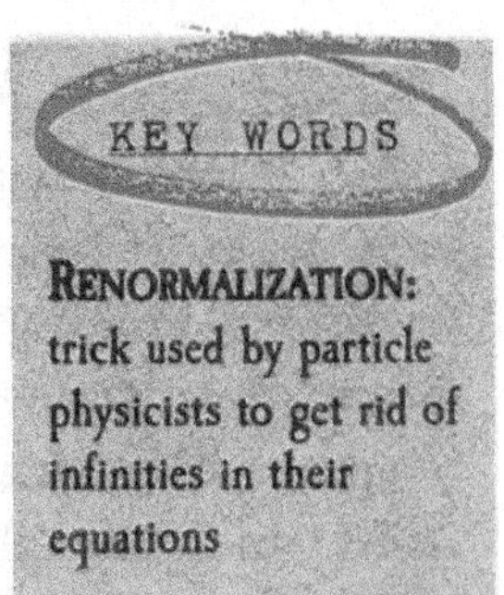

RENORMALIZATION: trick used by particle physicists to get rid of infinities in their equations

ZERO PROBLEMS

★ ***The main problem is that entities with zero volume allow infinite quantities to come into the equations.*** For example, the electric force associated with the charge on an electron is inversely proportional to the distance, from the electron, of the object being affected. The closer you are (if you are another charged particle), the bigger the force. If the electron has zero size, you can get to zero distance from it – and one divided by zero (let alone by zero squared) is infinity.

Gabriele Veneziano was one of the first string theorists

The birth of string theory

At the end of the 1960s new results were coming out of the particle accelerators, and different kinds of 'new' particle were being made there (and promptly decaying into the familiar particles). There were many suggestions as to what might be going on. A particularly intriguing one came from **Gabriele Veneziano,** in the form of a set of equations (a mathematical model) designed to describe the patterns made by the interactions. ***When physicists looked at the implications of these equations, they were surprised to realize that they didn't describe pointlike particles – but tiny lines, like bits of string.***

TANGLING WITH STRING

I can't get out!!

I can't get out!!

★ The first proper string theory came from Yoichiro Nambu, **who tried to describe particles in terms of spinning and vibrating lengths of string, each one only about one ten-thousandth of a billionth of a centimetre long. The idea was that one kind of string could vibrate in different ways, like different notes being played on a guitar string, to produce states corresponding to different particles. Some properties of the strings (like charge) were thought of as being stuck to the ends of the spinning strings.**

KEY WORDS

STRING THEORY: theory of the quantum world that describes particles such as electrons and quarks (and field quanta such as photons) in terms of the vibration of tiny loops of 'string'

EXTRA DIMENSIONS

★ To most people, the strangest thing about this kind of string is that it involves many more dimensions than the ones we are used to. ***In order to explain the properties of different kinds of 'particle' entities, the kind of string we are talking about has to vibrate in at least ten dimensions. This is straightforward to describe mathematically. But what does it mean physically? Where are the other six dimensions? And why don't we see them?***

a few people played with the equations that describe string

SUPERSTRING THEORY

★ The latest variation on string theory goes by the name of 'SUPERSTRING'. This is not just because physicists think it is better than ordinary string theory, but because it uses an idea called SUPERSYMMETRY, or SUSY. ***According to SUSY, there ought to be an exact correspondence between fermions and bosons. Because none of the known bosons correspond to the known fermions in the appropriate way, this calls for a new kind of fermion to partner every known kind of boson, and vice versa.*** These are termed SUPERSYMMETRIC PARTNERS, or SUSY PARTICLES. We can't see SUSY particles today because they only existed in the Big Bang, then quickly decayed. But they have been given names. The new bosons have names arrived at by adding an 's' to the appropriate fermion name (so the partner to the electron is the selectron), while the names of the new fermions end in 'ino' (such as photino). Supersymmetry explains many features of the particle world. So physicists at CERN, in Geneva, and at Fermilab, near Chicago, are now busy conducting high-energy experiments in search of the elusive SUSY particles.

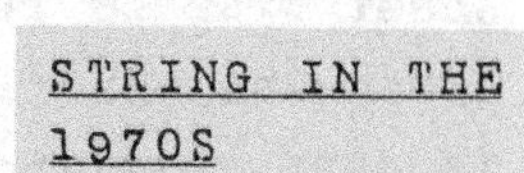

STRING IN THE 1970S

The first, accidental, version of string theory, which came out of Veneziano's equations, was readily explained in terms of quarks. So quark theory became the main line of research for particle theorists in the 1970s – and only a few people, mainly out of interest in the mathematics, played with the equations that describe string, which are not for the faint-hearted.

UNWANTED DIMENSIONS

* To get rid of unwanted dimensions, mathematicians use a trick called COMPACTIFICATION. If you view a garden hose from a very long way away, it looks like a one-dimensional line or thin string. But viewed more closely, it reveals itself to be a two-dimensional sheet wrapped round a third dimension. Similarly, a line is just a string of points, one after the other - but what looks like a single point on the line from a distance is really a little circle, or closed loop. This is compactification.

I thought this was a one-dimensional line, but it's really a garden hose

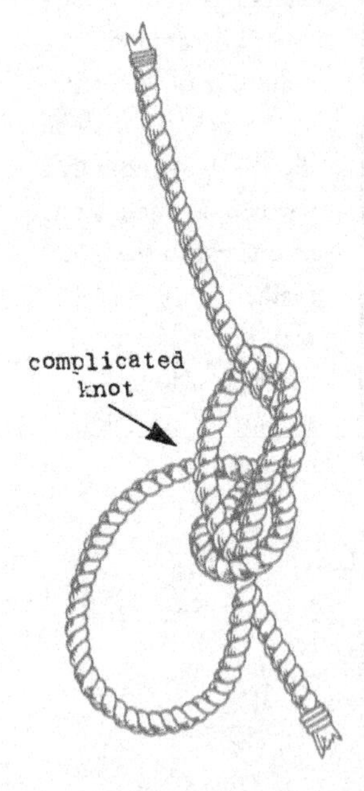

MULTIPLE ROLL-UP

* If you do the trick repeatedly, rolling up six or more dimensions in a more complicated knot, what you are left with looks – from a distance – like a sphere, or a line, or a little loop of string, depending on how you roll the dimensions up.

* *There are two important things about compactification. The first is that it will always work to hide dimensions from view – provided you apply it on a small enough scale, compared with the things you are actually measuring. The second is that*

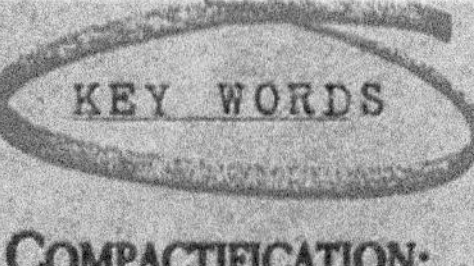

although the distances involved are small, they are not zero. So the problems with infinities do not arise, at least not in the best versions of string theory.

BOTH BOSONS AND FERMIONS

* Nambu's version of string theory worked, after a fashion. ***But it only described bosons, the field quanta that carry forces.*** It didn't describe fermions – which was embarrassing, since Nambu had set out with the intention of describing fermions, not bosons. ***But another physicist,* Pierre Ramond, *found a way to adapt Nambu's equations to describe both bosons and fermions in general terms, and this was developed into a theory describing bosons and fermions in terms of vibrating strings in ten dimensions.***

KEY WORDS

COMPACTIFICATION: trick used by mathematicians to 'roll up' unwanted dimensions and hide them from view

HOLDING ON TO STRING

Hardly anybody took much notice of these developments in the 1970s, because quark theory was causing so much excitement. The big buzz was QCD. Who needed string? By the early 1980s, just about the only two people still working at string theory seriously were **John Schwartz**, in the United States, and **Michael Green**, in England. They decided that if string theory was to be of any use to anyone, it would eventually have to describe all particles and all the forces of nature, gravity included. *In fact, it would have to be a quantum theory of gravity.*

QUANTUM GRAVITY

Quantum gravity had been a dream of physicists ever since the 1920s. ***Because gravity is so weak, quantum effects only become important for gravity on a tiny distance scale, known as the*** PLANCK LENGTH. This is virtually impossible to picture since it is only 10^{-33}cm or 0.000000 0000000000000000 0000000001cm (the nucleus of an atom is a hundred billion billion times bigger than the Planck length). ***If quantum theory applies to gravity, then on this scale spacetime itself is grainy, or granulated, so that the Universe can be thought of as a vast chessboard made up of tiny squares (or cubes) measuring just one Planck length across.***

SHRINKING STRING

*** Any thoughts of bringing gravity into the quantum fold were still just a dream in the early 1980s,** but Schwartz and Green knew they would take a step towards quantum gravity if they could shrink Nambu's string theory to the Planck length. At first, they did this with open strings, like Nambu's. Then they came up with another idea - closed loops of string.

NEAT LOOPS

* *Closed loops turned out to be a particularly neat idea – because different ripples running round the loop one way correspond to different kinds of bosons, while different ripples running round the loop the other way correspond to different kinds of fermions.*

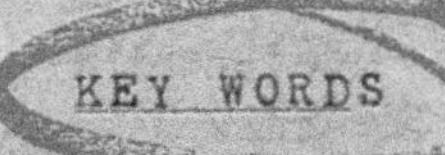

* But it was very difficult to make the properties of these 'particles' match up with the properties of known particles – even though the closed-loop versions of string theory did away with the infinities that plague theories like QED and did so in a pleasingly straightforward way. ***Among other difficulties, the string theories seemed to be predicting too many particles – not just too many individual particles, but too many varieties as well.***

STRING THEORY TAKES OFF

* ***One problem had been particularly worrying. It turned out that compactification only works if you start out with an odd number of dimensions in the first place. And ten is definitely not an odd number.*** But in spite of the difficulties, interest in string theory suddenly took off in the middle of the 1980s, when it turned out that one of the puzzling things in the theory that Schwartz and Green had been worrying over was actually just what they had been looking for.

KEY WORDS

PLANCK LENGTH: the quantum of length – the smallest length that has any meaning

Schwartz and Green

The American physicist John Schwartz was appointed to a professorship at Caltech in 1989. He has also worked in Paris and London, and at other American institutions. British physicist Michael Green has worked at several universities and research institutions in Britain and the United States, and at CERN in Geneva. In 1993 he took up a professorship at the University of Cambridge. It was almost solely thanks to Schwartz and Green that the idea of string theory was kept alive between the mid-1970s and the mid-1980s.

THINKING ABOUT GRAVITY

* To understand why string theory turned out to be of such importance to physicists, let's take another look at ideas about gravity. Einstein's theory of gravity is in essence a 'classical' theory, like Maxwell's theory of electromagnetism. It describes interactions between massive particles (which means any particles that have mass, not necessarily very heavy ones) in terms of the gravitational field, just as Maxwell's equations describe interactions between charged particles in terms of the electromagnetic field.

Belated recognition

In the 1960s nobody took much notice of Feynman's ideas about gravity, but in the 1990s his way of looking at gravity became very popular and successful – not least because of the connections with string theory that we are about to consider.

THE TROUBLE WITH GRAVITONS

* In QED, quantum theory developed this idea for electromagnetism, by describing the field in terms of quanta – in the form of bosons (or more specifically, photons) – that are exchanged between charged particles. *It is easy to see that a proper theory of quantum gravity would do the same sort of thing by describing interactions between massive particles in terms of the exchange*

of the appropriate field quanta – again bosons (in this case, gravitons).

* Any quantum theory of gravity would be bound to be more complicated than QED, because the equations tell us that ***gravitons, like gluons, can interact with each other (even though they each have zero mass) as well as with massive particles.*** But in principle it could be done, and this was the traditional route followed by most people who tried to find a quantum theory of gravity.

FEYNMAN AND GRAVITY

* There was one notable exception. **Richard Feynman**, who had invented quantum electrodynamics, became intrigued by gravity. In the early 1960s, even though he could not find a complete theory of quantum gravity, he was able to prove mathematically that ***if you started out with a quantum field theory involving the exchange of gravitons, then at an everyday level this would be indistinguishable from the general theory of relativity.***

Feynman explained superfluidity when very cold liquids flow without friction

Richard Feynman (1918–88)

Arguably the greatest physicist since Newton (certainly on a par with Einstein), Feynman achieved wider fame with his memoirs and when he pinpointed the cause of the Challenger disaster. But many people do not appreciate the scale of his contributions to physics. In addition to developing QED (the work for which he won the Nobel Prize), he made major contributions to our understanding of the weak interaction, QCD and gravity, each also worthy of a Nobel Prize, and almost as an aside explained a strange quantum phenomenon called superfluidity (when very cold liquids flow without friction). An inspiring teacher, he produced a series of textbooks still influential today.

TWO UNITS OF SPIN

* In the mid-1980s Schwartz and Green were trying to get their string theory to describe the known fermions and bosons. They expected it to be very difficult to include gravity, so first tried to make the particles predicted by their theory match up with things like photons and gluons, quarks and leptons. But their equations kept throwing out a different kind of particle - one with zero mass and spin 2. Eventually, the penny dropped. Their string theory was predicting the existence of gravitons. And that meant, according to Feynman, that it was predicting the general theory of relativity. They had arrived at a 'Theory of Everything'!

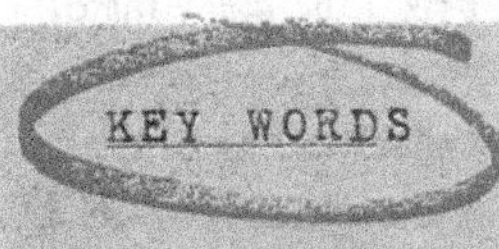

SPIN:
A quantum property which is subtly different from the way something like a spinning top, or the Earth, spins. In the quantum world, sometimes you have to rotate through 360 degrees *twice* to get back to where you started.

VARIATIONS ON THE STRING THEME

* Schwartz and Green had, in fact, discovered that there were only a handful of candidates for this Theory of Everything – ***just one particular family of variations on the string theme that worked in this way and had no infinities troubling it.***
* Before the mid-1980s, there had been lots of variations on the theme. Because there is no way to test these ideas directly

you can't test the equations, so you can dream up whatever you like

by experiment, you could dream up anything allowed by the maths and it might be a way of describing the real world.

MATHEMATICALLY CONVINCING

* It's a bit like what would happen if it was impossible to test theories of gravity by looking at things like the orbits of the planets around the Sun. You might guess that there was an inverse-square law, or an inverse-cube law, or some other kind of law that you could describe mathematically. But you wouldn't be able to prove which version was the right one. And nobody would take much notice of such a vague theory.

* ***So when it turned out that only one kind of string theory was free from infinities, it was as if mathematics, not experiments, was saying that the only true law of gravity would have to be an inverse-square law. And then people had to start taking it seriously.***

Zero mass, spin 2

The key properties of the graviton are that it has no mass, and has two units of a property quantum physicists call SPIN. This is not like the spin of a spinning top, or the spin of the Earth on its axis. It's just one of those quantum labels you have to get used to. ***Feynman had shown that if you had gravitons with zero mass and spin 2, then you would automatically get the general theory of relativity.***

the graviton has no mass and two units of spin

THE POWER OF MATHEMATICS

* It may seem strange that mathematics alone can be used to tell us which sort of string theory applies to the real Universe, without ever doing experiments. But in fact the analogy with a theory of gravity based on an inverse-cube law really works, at this level.

THE ANTHROPIC PRINCIPLE

According to the anthropic principle, the very fact that we exist tells us what the Universe at large is like. For example, life forms like us depend on the chemistry of carbon. So carbon must be made in the Universe. In the 1950s, Fred Hoyle used this requirement of our existence to predict that certain nuclear reactions would be able to make carbon inside stars. Experiments in laboratories here on Earth then confirmed that those reactions are possible.

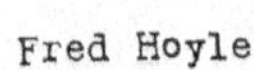

Fred Hoyle

it's all there in the equations

NO STABLE ORBITS...

* *If you try to describe (mathematically) an imaginary universe in which gravity obeys an inverse-cube law, it turns out that there are no stable orbits.*

* In our Universe, with an inverse-square law of gravity, if a planet like the Earth is struck by a meteorite and nudged a tiny bit closer to or further away from the Sun, then it automatically moves back towards its old trajectory. You don't have to do

experiments to find this out – it is built into the equations.

* But the equations also tell us that in a universe obeying an inverse-cube law of gravity, a planet that moved a little closer to the Sun, for whatever reason, would keep falling inwards towards the Sun. And a planet that moved a little further out, for whatever reason, would keep spiralling out.

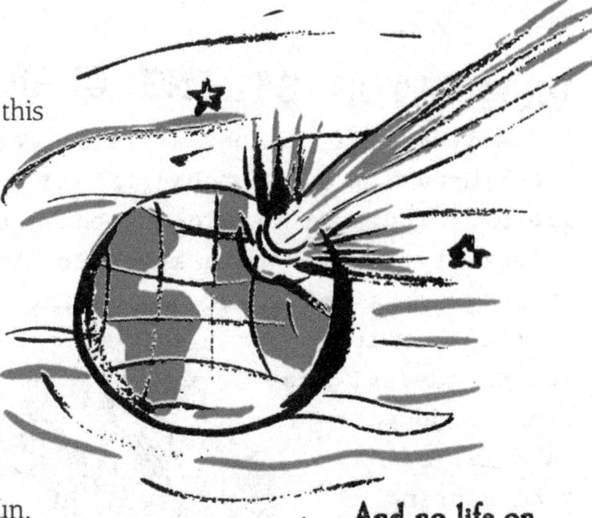

an inverse-square law of gravity tells us that the Earth cannot be knocked out of its orbit

NO SOLAR SYSTEM...

* At a more subtle level, if you follow the mathematics of curved spacetime through in the way Einstein did in his general theory of relativity, the equations tell you that an inverse-square law of gravity is the natural law ***only if there are three dimensions of space.*** So if a universe existed with more or less space dimensions than three, there would be no stable planetary orbits and no planetary system like our Solar System. So, presumably, no life as we know it.

life as we know it

...And no life on Earth

Some people argue that this is why we see that the extra dimensions of space have compactified to leave only three 'visible' on the large scale. Other universes might exist, but there would be nobody in them to notice. So the very fact that we are around to wonder about these things means that we must live in a Universe with three space dimensions. This sort of argument can be applied to other features of the visible Universe and to other laws of physics. It is sometimes called the ANTHROPIC PRINCIPLE, and people still argue about whether it is more than a mere tautology.

PRINCETON STRING QUARTET

* One team that enthusiastically followed up the breakthrough made by Schwartz and Green was a group of four mathematicians based at Princeton University. They became known to their colleagues as the Princeton String Quartet - which gives you an idea of the kind of jokes mathematicians find amusing.

Princeton String Quartet

MORE DIMENSIONS STILL

The Princeton String Quartet's chief contribution to string theory was to write down the equations describing closed loops of string, using a different mathematical approach. You need a PhD to understand the mathematics, but fortunately for us their variation on the theme corresponds to a very neat physical picture of what is going on (provided you are willing to accept a few more extra dimensions of space).

HETEROTIC STRING

* The quartet's version of string theory glories in the name of 'HETEROTIC STRING'. The name comes from the Greek root 'hetero' (as in heterosexual), implying two different varieties of something. ***It is apt, because in heterotic string theory the kind of vibrations associated with fermions require ten dimensions, whereas those associated with bosons require no less than 26 dimensions. The ten-dimensional vibrations run one way around the loop of string, while the 26-dimensional vibrations run the other way.***

now smear the properties along the string

* It isn't really surprising that the bosonic vibrations require more dimensions, because there is a much greater variety of bosons (which range from photons to the W and Z particles, gluons, and gravitons) than there is of fermions. Admittedly, fermions include six quarks and six leptons, but there are only two basic 'body plans' for each.

FOUR OUT OF TWENTY-SIX

* The equations that describe this richness of particles are interpreted as implying that 16 of the 26 dimensions have compactified as a set, providing the framework for the bosonic vibrations. The ten remaining dimensions provide scope for the fermionic vibrations, with six of these ten dimensions compactifying in a different way – leaving the familiar four dimensions of spacetime.

Smeared along the string

Properties like electric charge are also properly described by heterotic string theory, and can be pictured as being smeared out around the entire loop of string.

there is no way to test string theory

DOPPELGANGER SCENARIO

* There is one intriguing curiosity about heterotic string theory. It doesn't just describe fermions and bosons in one complete package. It does so *twice*. The equations allow for two complete sets of everything - everything known to particle physics is described by one half of the equations, and the other half of the equations seems to describe another complete set of particles and fields.

Testing string

As already mentioned, there is no immediately available way to test string theory definitively. However, there is one test that would go some way towards proving its validity. ***Most versions of the theory predict the existence of another kind of Z particle, as well as the one we already know. If experimenters found this particle, it would be powerful evidence in support of the theory.*** But if they don't find it, that won't settle the issue – since absence of evidence is not the same thing as evidence of absence.

TWO OF EVERYTHING

* Some quantum theorists think this means that at the birth of the Universe, in the first split-second of the Big Bang, two completely independent sets of particles and fields were created. These duplicate worlds would only interact at all through gravity. ***According to this picture, there could be another universe of stars, planets and even people occupying the same four-dimensional spacetime as our Universe. This hypothetical ghostly duplicate universe is called the*** 'SHADOW UNIVERSE'.

ELUSIVE SHADOWS

* The idea is that there would be shadow particles (the equivalent of quarks, photons and all the rest) making the shadow stars and planets, all for ever invisible to us. Except for the fact that we would notice its gravitational pull, a shadow star or planet could pass right through the Earth, without either of them being affected. Unfortunately (or fortunately?), short of a collision of this kind between a shadow object and ourselves, there seems to be no way to test the idea. *

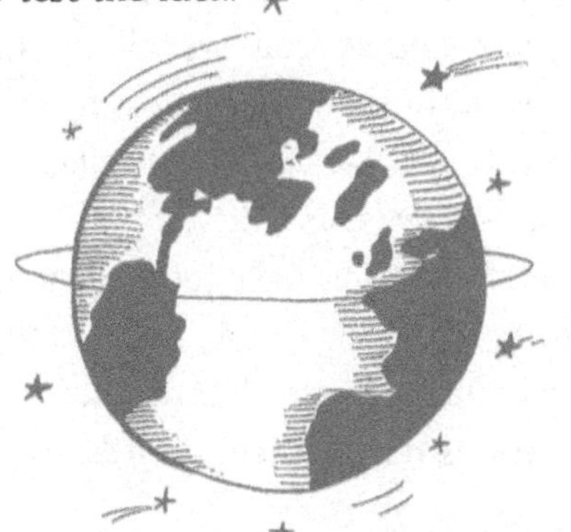

GREAT LEAP FORWARD

In the second half of the 1990s, despite all the difficulties, string theory took a great leap forward when the theorists found a way to get rid of the problem of having an even number of dimensions. Their suggestion looks outrageously obvious, almost silly. But you have to remember that it is backed up by very sound mathematical physics. It isn't just wishful thinking.

INTO THE 21ST CENTURY

* The great idea of the late 1990s was that, instead of thinking in terms of one-dimensional strings, we ought to be adding in another dimension - so as to produce two-dimensional sheets, usually referred to as MEMBRANES.

Membranes or magic

Remember how in basic string theory (see page 140) a hosepipe is treated as a two-dimensional sheet (membrane) rolled up to give the appearance of a line? In membrane theory, although you are dealing with 11 dimensions (at one level, 27 at the other), one of them is immediately rolled up – so that the membrane behaves like a ten-dimensional string. The resulting mathematical package is called M-THEORY because the M stands for membrane, mystery or magic – the experts say you can choose whichever name you like!

it would be nice to have one single theory

bringing M-Theory out of the hat

THE CASE FOR M-THEORY

* The reason why people take this idea seriously is that there are still, even after the breakthrough made by Schwartz and Green in the 1980s, several different versions of basic string theory – and it would be nice to have just one single theory, a unique Theory of Everything. ***It turns out, however, that all these variations on the theme can be expressed***

in terms of a single M-theory. This is rather like the way electromagnetism and the weak interaction turned out to be different facets of a single mathematical package, the electroweak theory.

THE END IS NOT NIGH

★ ***As we enter the 21st century, many physicists are convinced that M-theory provides the ultimate description of the way the Universe works at the deepest level, down on the Planck scale.*** But this does not mean the end of physics, because they are just beginning to realize how much remains to be explained at higher levels – when huge numbers of particles are involved, working together to produce the complexity of a breaking ocean wave, the pattern of orbital motions of the planets in the Solar System, the complexity of a living creature, or the workings of the weather.

METAPHYSICAL PHYSICS

Today's buzzwords are chaos and complexity. These have introduced new areas of study where there are still more questions than answers, and which are likely to keep physicists busy for many decades to come. The topics with which the newest physics is concerned were first broached more than a hundred years ago, but they seemed so strange and incomprehensible that they were largely ignored until the late 20th century. Then the concept of chaos was revived and became a major branch of physics, largely thanks to the availability of powerful computers capable of probing the implications of the new science.

physics has a lot to explain about the interactions of large numbers of particles

CHAPTER 5

BACK TO THE FUTURE

I'm sowing the seeds of the new physics

* The seeds of the newest physics were sown in the mid-1880s, when a maths professor in Stockholm announced a competition to honour the sixtieth birthday of the king of Sweden and Norway, Oscar II, due in 1889. A generous prize was offered for the best piece of work on one of four specified puzzles, one being the question of whether or not the Solar System is stable - a topic that, in a sense, dated back to Isaac Newton's work on gravity.

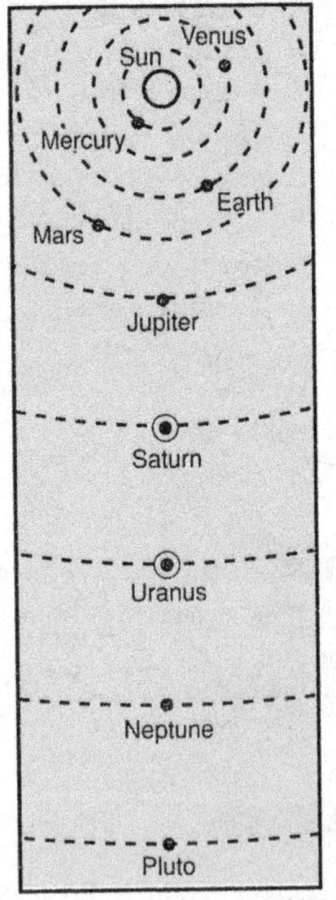

solar system

NEWTON AND GRAVITY

* Newton's inverse-square law of gravity describes the orbit of a single planet around a single star perfectly. It affirms that the orbit is an ellipse, exactly matching both the observations made by **Tycho Brahe** and explained by **Johannes Kepler** (see page 7) and all the observations made ever since. ***But if you have two or more planets orbiting the same star, they tug at each other, so that their orbits are no longer perfect Keplerian ellipses.***

Isaac Newton

UNSOLVABLE EQUATIONS

* Using his own equations of gravity and orbital motion, Newton could not find a way to describe the resulting orbits accurately. This wasn't because he was incompetent, or because it was impossible without computers. ***It's simply that the equations cannot be solved.***

* An elliptical orbit can be described by a very simple DETERMINISTIC FORMULA – called an ANALYTICAL SOLUTION to the equations of motion – which can be used to tell us where a planet will be in its orbit at any time in the future, and where it was at any time in the past. ***But if three or more objects are involved, all affecting each other by gravity, there are no analytical solutions to the equations. In fact, there's no option but to laboriously work through the complete calculation, starting from the equations of motion and the law of gravity, for each individual case.***

the orbit is an ellipse, observes Tycho Brahe

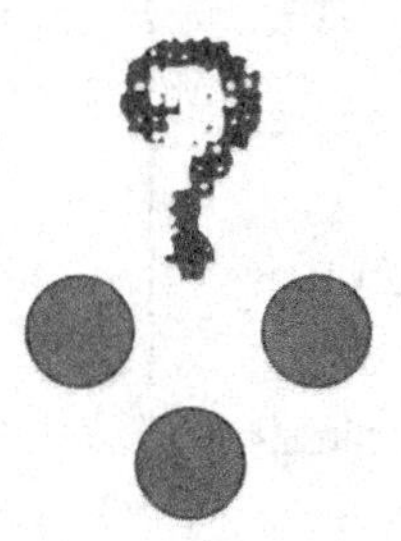

how can we solve the three-body problem?

DETERMINISTIC FORMULA:
a simple equation that tells you exactly how something will change (for example, the equation of an ellipse tells you how a planet moves round the Sun)

ANALYTICAL SOLUTION:
another name for a deterministic formula

TAKING IT STEP BY STEP

The only way to 'solve' the three-body problem is step by step. Pretend two of the bodies do not move and calculate the orbit of the third one. Then move it a tiny bit forward in its orbit. Now hold that body still, plus one of the others, and calculate a tiny move in the orbit of the remaining one. Repeat for the third body and continue *ad infinitum*. (This is all done on a computer of course, not in the real world.)

CELESTIAL PREDICTABILITY

Newton believed that God controlled the Universe

* What troubled Newton was that if a planet were tugged even slightly out of orbit by the gravitational influence of another planet, it might not go back where it belonged. He feared that the Earth, or another planet, might one day start to spiral in towards the Sun or out into the depths of space. However, he was a deeply religious man and consoled himself (and his readers) by suggesting that God would intervene to nudge the planet back again into its proper orbit.

God will not be necessary

When Laplace's work was published in book form, Napoleon commented that there was no mention of God anywhere in its pages. To which, Laplace proudly replied **'I have no need of that hypothesis.'**

COMPLEX TIMING

* In the 1780s the French mathematician and astronomer Pierre Laplace took a careful look at the problem that had worried Newton, concentrating on the orbits of Jupiter and Saturn - the two largest planets in the Solar System, with the strongest gravitational influence after the Sun itself.

* He found that the orbit of Jupiter was expanding slightly, while the orbit of Saturn was shrinking slightly. He linked

Jupiter

these changes to the gravitational influence of the two planets on each other. This operates in a particular rhythm: Saturn makes two orbits of the Sun while Jupiter journeys around the Sun five times, so they are close to each other roughly every 59 years. Using Newton's laws, Laplace calculated that the change in the orbits would reverse after about 929 years – with the orbit of Jupiter shrinking and the orbit of Saturn expanding – and this cycle would be repeated roughly every 929 years.

REGULAR AS A CLOCK

Saturn

* Laplace was so impressed by the habits of Jupiter and Saturn that he wrote: ***'The irregularities of these two planets appeared formerly to be inexplicable by the law of universal gravitation – they now form one of its most striking proofs.'*** And he went on to show that the same kind of stability applied, as far as he could tell, to the orbits of all the planets – so that the whole Solar System was as stable and regular as a clock.

Pierre Laplace

Pierre Laplace (1749–1827)

The son of a magistrate in a small town in Normandy, Laplace studied under the Benedictines at a local college, then moved on to Caen and from there to Paris, where he became a professor of mathematics in 1768. One of the few members of the Paris elite who held government posts both before and after the French Revolution, he was a member of the committee that introduced the metric system of weights and measures. Laplace served in the Senate and briefly as Minister of the Interior, but voted for the restoration of the monarchy in 1814. This didn't help his prospects when Napoleon returned to power, but he got his reward when Louis XVIII made him a marquis.

THE QUESTION OF STABILITY

* Why the stability of the Solar System figured in the competition honouring Oscar II wasn't because mathematicians thought Laplace had been wrong. Quite the reverse. In 1858 Peter Dirichlet had worked out a new method of solving differential equations - which, he claimed, could prove that the Solar System is stable. Dirichlet died in 1859, and nobody had succeeded in reconstructing his calculation. Hence the question included in the competition.

is the Solar System stable?

Peter Dirichlet (1805–59)

German mathematician who became a professor in Berlin in 1828, then moved to a similar post in Göttingen in 1855. His predecessor there was the great mathematician Karl Gauss, and one of his most notable contributions to mathematics was to clarify Gauss's work and make it more accessible. Dirichlet set great store by conceptual clarity in mathematics. Although shy and modest, he had a major influence on the way the subject developed.

DAWN OF A NEW ERA

* The prize was won by a French mathematician, **Henri Poincaré**, for a paper that didn't really answer the question. But it was clearly a profound new development in mathematics – 'of such importance,' the judges wrote to the king, 'that its publication will inaugurate a new era in the history of celestial mechanics'.

* ***Poincaré had come up with a new way of looking at problems like orbital motion that did not depend on working out the orbit (or whatever it was you were interested in) step by step***. He pointed out that if any kind of system starts out from a particular state and does something

complicated, then gets back to exactly the same state it started in, then it is bound to repeat its whole complicated pattern of behaviour.

Poincaré had found a new way of looking at the complexities of the Solar System

CYCLICAL BEHAVIOUR

* Take the (very simple) case of a planet orbiting around the Sun. If you notice that at a particular point in its orbit the planet is moving in a certain direction at a certain speed and later see the same planet in the same place travelling in the same direction at the same speed, then you know it must be in an orbit that will repeat indefinitely. ***Poincaré realized that there is therefore no need to carry out a calculation of every step of its orbital motion, or even to discover what it does in the rest of its orbit.*** You know that whatever it was it did, it got back to exactly where it started.

BEGINNING

I'm back where I started

KEY WORDS

DIFFERENTIAL EQUATION: equation that describes how something affects the rate of change (often over a period of time)

Henri Poincaré

Henri Poincaré (1854–1912)

Poincaré qualified as a mining engineer, but he was always keenly interested in mathematics and in 1881 became a lecturer at the Faculty of Sciences of the University of Paris, where he spent the rest of his career. He made many contributions to many areas of science, including the understanding of probability theory and planetary orbits. In the 1890s, when Einstein was still at school, Poincaré thought and wrote about the implications of relative motion. As early as 1900 he pointed out that Maxwell's equations require that nothing can travel faster than light, but never followed this up with a complete theory of relativity.

WAITING FOR EVER

* If you have a box divided in two, with gas in one half and empty space in the other, when you pull the partition out the gas spreads to fill the box. You never see all the gas congregating at one end. But Poincaré argued that if you were able to wait long enough you could see this happen, because the atoms of gas bouncing off the walls of the box must eventually pass through every state allowed by the laws of physics - including a stage where all the gas is at one end of the box.

MORE ZEROS THAN STARS

* ***The key words here are 'long enough' and 'eventually'.*** A small box of gas might contain 10^{22} atoms (a 1 followed by 22 zeros, or 10,000 billion billion atoms), and there is an enormous number of different ways in which so many atoms can be arranged. The time it would take for so many particles to pass through every possible state is much greater than the age of the Universe. ***Typical Poincaré cycle times have more zeros after the 1 than there are stars in all the known galaxies in the Universe put together – numbers so big that it hardly matters whether you are counting in seconds, hours, or even years.***

* ***Poincaré had found the key to chaos, but nobody used it to unlock the box.*** All the mathematical tools needed to develop chaos theory were in the hands of scientists at the beginning of the 20th century, but they did nothing about it for half a century. ***This turned out to be one of the biggest oversights in science.***

IMPROBABLE ODDS

Turn Poincaré's numbers around and they represent the odds against any particular state occurring in any particular second, or hour, or year during which you are watching the box. The point is that there are billions and billions of states corresponding to gas filling the box, so you are quite likely to see that. Whereas there are vastly fewer states corresponding to a box exactly half full of gas, so you are very unlikely to witness such a phenomenon.

watch the orbits

THE THREE-BODY PROBLEM

*** If you have two bodies (planets, for example) with equal mass alone in space, they will orbit around each other in exactly predictable elliptical orbits. Now extend this picture by adding in a single tiny particle of dust, so small that its gravitational influence on the two planets can be ignored. What is the orbit that this dust particle follows around the two larger objects?**

Poincaré and orbital dynamics

Poincaré thought his approach could, in a much more rigorous mathematical form, be used to prove the stability of the Solar System. But he was wrong. Not surprisingly perhaps, because even the simplest version of the classic 'three-body problem' shows just how complicated orbital dynamics is.

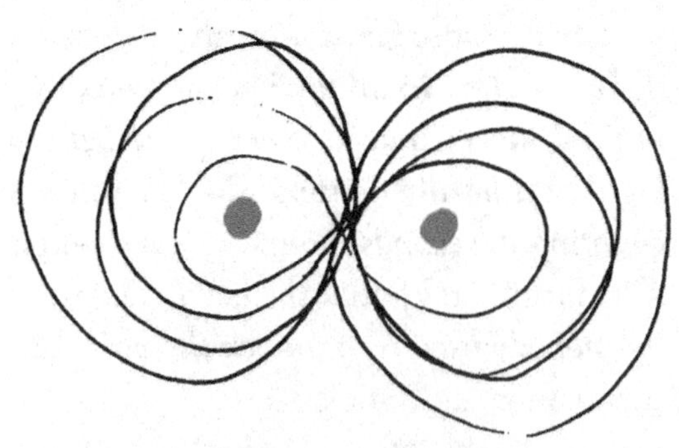

the orbit of the tiny particle around the larger particles is completely random

INTERMINABLE CALCULATIONS

* There is just no simple answer to this question, unless the dust particle is far enough away from the two planets for it to be able to orbit around the pair of them. Otherwise, it sometimes follows a complicated looping orbit, like a tangled ball of string, around one of the objects.

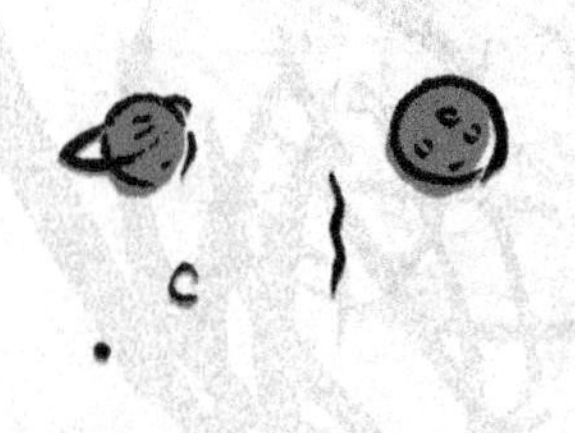

Then, at other times, it crosses over and orbits in the same sort of complicated way around the other planet. And sometimes it follows a confused kind of figure-of-eight pattern, orbiting around each of the planets in turn.

* The only way to work out the path of the orbit is to trace it step by step using a computer (or, much more laboriously, a pencil and paper). At each instant, you can work out all the forces involved and see how the dust particle will move a tiny bit. ***But all the forces immediately change as a result, because the dust particle is now in a different position relative to the two planets (which have also moved in their own orbits). So you then have to calculate all the forces again and work through the next step.***

NO RETURN TO SQUARE ONE

The crucial thing about the three-body problem is that you ***never*** get back to a point where the particle is moving in the same direction at the same speed as before, with the two planets also in their original places and moving at the same speeds and in the same directions as before. ***The orbit does not repeat – and there is no simple formula describing it, like the formula that describes an ellipse.***

Tiny changes can make big differences

The ultimate form of the orbit followed by the dust particle in the three-body problem depends on where it starts from and at what velocity. If you were to make a tiny change to the Earth's speed or direction through space, it would still follow almost exactly the same orbit as before – because its motion is almost entirely dominated by the gravity of the Sun. ***But if you give the dust particle a tiny nudge, altering its speed or direction ever so slightly, at some points in its orbit that would be enough to make it veer off onto a completely different, though equally complicated, looping orbit.***

COMETS AND CHAOS THEORY

* There are, in fact, things in our Solar System that behave very much like the dust particle in the three-body problem. And, like the dust particle in the problem, they are small objects - things like comets and asteroids, much smaller than planets.

A CRITICAL NUDGE

* Comets are basically chunks of icy material, a few kilometres or tens of kilometres across. Many of these icy chunks can be seen in what seem to be stable orbits, in the outer part of the Solar

comets are very sensitive to the gravitational influence of the planets

the Sun is the dominant body in the Solar System

System, near the giant planet Jupiter. But computer calculations tell us that a tiny nudge at certain points in their orbits could switch objects such as comets onto a completely different trajectory, perhaps tumbling past the Earth and in towards the Sun. Just such a nudge is provided from time to time, by the gravity of Jupiter, or Saturn, or something else.

* ***What matters is that these orbits are extremely sensitive to small changes, if those changes are applied at certain times. And this 'sensitivity to initial conditions' is what*** <u>CHAOS THEORY</u> ***is all about.***

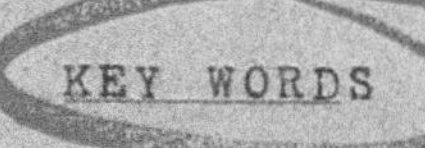

Chaos:
Term used to describe the behaviour of systems that are very sensitive to initial conditions. A tiny change in the starting conditions makes a big difference to where you end up. Chaotic systems are essentially unpredictable, because you can never specify the starting conditions accurately.

Planetary influences

In the Solar System, the Sun is so much bigger than any of the planets that its gravity dominates and is by far the most important factor in deciding what orbits the planets follow. But it isn't the only factor. The planets themselves have an effect on each other.

I didn't touch it!

there are factors in the Solar System apart from the Sun that affect planetary orbits

FROM RAINDROPS TO CASCADES

* It's easy to appreciate how a tiny difference at the outset can have a big influence on how or where you end up. Think of a drop of rain falling on a high mountain range such as the Rocky Mountains of North America. Rainfall that lands on one side of the mountains will flow westwards and end up in the Pacific, while rain landing on the other side is likely to end up in the Atlantic or the Gulf of Mexico.

sometimes a tiny cause can have a big effect

THE IMPACT OF INITIAL CONDITIONS

'A very small cause that escapes our notice determines a considerable effect that we cannot fail to see, and then we say that the effect is due to chance. If we knew exactly the laws of nature and the situation of the Universe at the initial moment, we could predict exactly the situation of that same Universe at a succeeding moment. But even if it were the case that the natural laws had no longer any secret for us, we could still only know the initial situation approximately. If that enabled us to predict the succeeding situation with the same approximation, that is all we require, and we should say that the phenomenon had been predicted, that it is governed by laws. But it is not always so; ***it may happen that small differences in the initial conditions produce very great ones in the final phenomena. A small error in the former will produce an enormous error in the latter. Prediction becomes impossible, and we have the fortuitous phenomenon.***'
Henri Poincaré

CRITICAL POINT

* Somewhere high in the mountains, there must be a point or a line where a tiny difference (maybe only a millimetre or so) in where the raindrop lands will make a difference of thousands of kilometres in where it ends up.
* Which ocean it ends up in may not matter to the raindrop, but it certainly matters to us. ***And sometimes such a tiny initial difference can trigger a cascade of events with dramatic consequences.***

'It may happen that small differences in the initial conditions produce very great ones in the final phenomena. A small error in the former will produce an enormous error in the latter. Prediction becomes impossible and we have the fortuitous phenomenon.'

HENRI POINCARE

CHAINS OF EVENTS

* Domino effects of this kind are hardly new to human experience. Indeed, there is a well-known old rhyme that makes the point forcefully:

> For the want of a nail, a shoe was lost;
> For the want of a shoe, a horse was lost;
> For the want of a horse, a rider was lost;
> For the want of a rider, a battle was lost;
> For the want of a battle, the kingdom was lost.

What is new is the realization that this kind of magnification of small events, culminating in major influences, applies in the world of science. And this kind of phenomenon can be described, by mathematics, in a scientific way. Chaos may not be predictable but it is scientific.

IMPRECISE CONDITIONS

* It wouldn't matter that systems are sometimes sensitive to initial conditions if you could always specify the initial conditions completely accurately. But even apart from the question of quantum uncertainty, sometimes it is impossible to specify even such a simple thing as the position of an object with complete precision.

> **NOTHING UNCERTAIN**
>
> 'Assume an intelligence that at a given moment knows all the forces that animate nature as well as the momentary positions of all things of which the Universe consists, and further that it is sufficiently powerful to perform a calculation based on these data. It would then include in the same formulation the motions of the largest bodies in the Universe and those of the smallest atoms. To it, nothing would be uncertain. Both future and past would be present before its eyes.'
> *Pierre Laplace*

RATIONAL NUMBERS

* Suppose you are only trying to specify the position of an object at a particular point on a line, without worrying about its position in three-dimensional space. Label one end of the line '0' and the other end '1', and divide it into any units you care to use. It is then easy to specify the position of ***some*** points on the line, using fractions such as $^{1}/_{2}$ or $^{3}/_{7}$ or even $^{2573}/_{6937}$.

* These fractions are ratios of one number to

another, which is why they are called 'rational' numbers. You can find a rational number that is as close as you like to any point on the line – even the ancient Greeks knew this.

IRRATIONAL NUMBERS

* But there are points on the line that can never be labelled precisely by any fraction that is a ratio of two ordinary numbers. These points correspond to irrational numbers, which are expressed as an infinitely long string of numbers after the decimal point, with no repeating sequence to it. The word 'irrational' doesn't mean they are illogical – just that they are not ratios. ***The lack of a repeating pattern in these numbers is very similar to the way the orbit of the dust particle in the three-body problem never repeats itself.***

Asking too much

If the position on the line you need to specify corresponds to an irrational number, then to spell it out precisely you would need a decimal expression with an infinite number of numbers in it. This would require an ***infinite*** amount of information – so no computer, or intelligence of the kind envisaged by Laplace, could work it out precisely. ***This is where Laplace's vision of a perfectly predictable Universe falls down. The only computer powerful enough to describe the behaviour of the entire Universe would be the entire Universe itself.***

CHAPTER 6

TOMORROW'S PHYSICS

*** The first person to notice chaos at work in a practical context was not a physicist but an American meteorologist,** Edward Lorenz. **The effects of chaos suddenly confronted him while he was working with one of the first computers used to tackle the problems of weather forecasting, in the early 1960s.**

Science in the 20th century

The reason why scientists didn't turn their attention to phenomena such as chaos sooner was that they were busy solving all the things that could be solved using Newton's approach to science. They achieved so much, so quickly, by concentrating on equations amenable to analytical solutions that there was no incentive to tackle the much messier situation of equations they couldn't solve analytically – until the importance of such situations was rammed home by a series of discoveries in the second half of the 20th century.

PREDICTING THE WEATHER

* The idea behind this kind of weather forecasting is that the laws of physics governing how winds move around the globe and how the temperature changes, and so on, can be described by sets of equations that can be solved using a computer. The computer simulations are fed with numbers corresponding to things like the temperature and pressure of the atmosphere at different points – maybe on

a tiny change in where you start makes a big difference to where you end up

a grid with points separated by 100km over the surface of the Earth, and at 100km intervals up through the atmosphere. Then you plug the numbers into the equations and run the computer to calculate how the weather will change.

SURPRISE FORECAST

* One day, Lorenz decided to extend an earlier run. To save time, he started in the middle of the old run, typing in the numbers corresponding to the weather taken from the printouts of the first run. He assumed that the computer would duplicate the second half of the old forecast, then extend it. But to his surprise, it didn't. ***Instead, the forecast weather gradually diverged from the previous run, until there was no resemblance at all between the two forecasts.***

the computer's forecast could change drastically with only tiny changes in the initial data

FEEDING IN THE DATA

The computer Lorenz was using, back in 1961, was much more primitive than modern supercomputers, and he wasn't trying to predict the real weather but merely looking at how the computer handled some of the simplest equations that would be needed in real forecasting. It seemed to work. When sensible numbers were fed into the computer, plausible 'forecasts' of things like temperature came out. You could start the machine off in a certain way, and get numbers out corresponding to things like the temperature variation day by day, for weeks ahead – as far as you wanted to run it. Or so everyone assumed.

THE BUTTERFLY EFFECT

* Lorenz's discovery that weather systems and forecasts suffer from chaos explains why accurate forecasts today are limited to a few days ahead (at most a week or two), and will *never* get any better. No matter how good your computer is, the effects of chaos begin to show up after about that time.

DISTANT FLUTTER

* Forecasters can see this by feeding very slightly different numbers into their computers and running a forecast again. For the first few days, the alternative forecasts will be almost identical; but after a week they look very different (unless the weather happens to be in an unusually stable pattern). No matter how accurate the equations are, and no matter how good the data you feed in are, after a week tiny effects that are too small for us to notice can exert a significant influence on the results. Lorenz called this 'THE BUTTERFLY EFFECT'. ***He imagined that a butterfly fluttering its wings in the jungle***

The discovery of chaos

Why Lorenz's computer presented him with such a divergent forecast was that Lorenz had rounded off one of the numbers he typed into the computer. Numbers used in the computer were kept to six decimal places – in this case, 0.506127. When he started his second run, Lorenz omitted the last three digits and typed in 0.506. ***This tiny difference was enough to drastically alter the forecast that came out of the machine.***

the figures just don't add up!

of Brazil could set up little eddies in the atmosphere that might change the weather in London months later.

CHAOS NAMED

* One reason why Lorenz's discovery didn't change the course of science immediately was that it was published in the *Journal of the Atmospheric Sciences*. The meteorologists who read it seem not to have realized that it had universal significance – and the physicists and mathematicians who might have appreciated the importance of the discovery didn't read weather journals. ***The person who gave chaos its name was indeed a mathematician,* James Yorke, *who came across Lorenz's paper in the early 1970s and realized that it gave a physical basis for a whole host of things that mathematicians had been playing with in the 1960s.***

IGNORE THE EQUATIONS

In traditional physics the equation is king. You start with equations (like Newton's laws of motion) and use them to calculate things. The equation is all you need. But chaos is different. James Yorke said that, 'Sometimes you can write down the equations of motion and sometimes you can't. Our approach is to ignore the equations.'

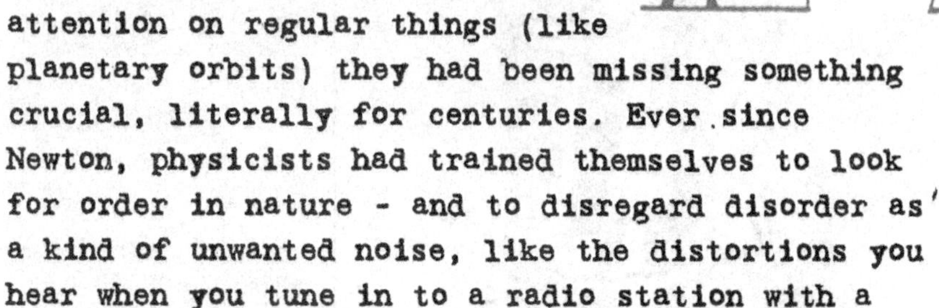

TUNING IN TO CHAOS

* When Yorke and other mathematicians publicized Lorenz's discovery, physicists began to see that chaos was all around them and that by concentrating their attention on regular things (like planetary orbits) they had been missing something crucial, literally for centuries. Ever since Newton, physicists had trained themselves to look for order in nature - and to disregard disorder as a kind of unwanted noise, like the distortions you hear when you tune in to a radio station with a weak signal. Now, they began to realize that the noise might be more important than the signal.

do it yourself chaos

Just like Lorenz

By the way, should you use the same starting number on your calculator and on your desktop computer or another calculator, don't be surprised if you end up with two different sequences of numbers. This is because the machines may round off the numbers in slightly different ways – ***just like the rounding-off process that led Lorenz to discover chaos in the first place.***

DIY CHAOS

* You can see chaos at work mathematically using a pocket calculator, or program your computer to do all the donkey work for you. Start with a simple mathematical expression, $(2x^2 - 1)$. You can

doing the same calculation on different computers will give different answers

work out a value for this expression, for any number you choose (x), by squaring x, doubling the result and then subtracting 1.
* Now do the same with the number you are left with, and repeat the whole process again and again (this is called ITERATION). You might expect that if you start out with two very similar numbers, after a few iterations you will end up with two similar numbers. But you don't.
* Try it with 0.51234, and make a note of each number you get at each step in the iteration. Now try it again, starting with 0.51235, again writing down the numbers at each step of the iteration. ***After about 50 iterations, the two sequences of numbers will be completely different. This is chaos. A simple deterministic formula leads to very different places from very similar starting conditions.***

ITERATION

Iteration is a mathematical process that goes round and round in a loop, starting out with some input (usually just a number), then carrying out a series of mathematical manipulations, and coming up with an output (another number) which is then used as the input to the next run around the cycle.

iteration is a process that goes round and round in a loop

CHAOS AT WORK

* You can see chaos at work, physically, in executive toys based on the principle of a double pendulum. A double pendulum is essentially two equal rods joined by a hinge in the middle. The double pendulum is set swinging by a motor that gives it a little nudge from time to time, or by being given a poke with your finger. Depending on just how it is nudged, the pendulum behaves in very different ways.

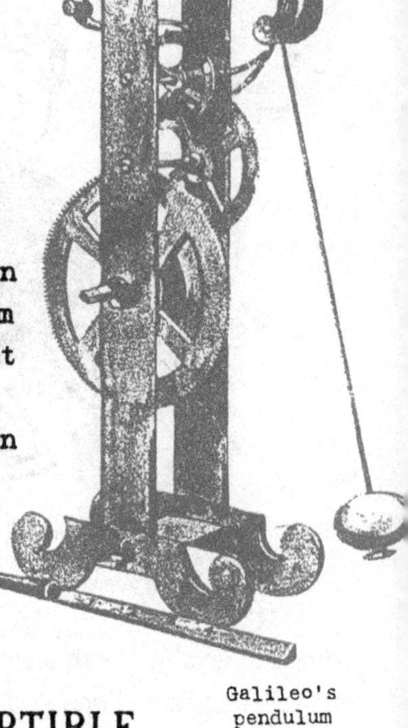

Galileo's pendulum

What is chaos?

To a scientist, chaos is not the kind of incomprehensible and messy situation that the word means to most people. Instead, it is a special kind of messy behaviour that results from entirely predictable physical laws, such as the ones formulated by Newton. ***It is unpredictable behaviour obeying predictable laws.***

IMPERCEPTIBLE DIFFERENCES

* Sometimes it swings as one piece, like a clock pendulum. Sometimes the bottom part swings wildly to and fro, while the top half hardly moves. And sometimes the

this toy is completely chaotic

bottom hardly moves, and the top hardly moves, but the joint in the middle part of the pendulum swings wildly to and fro. The pendulum seems to switch from one kind of oscillation to another for no reason, even though it has been given apparently identical nudges.

* ***The crucial point is that the nudges are very slightly different to each other, and even though these differences are too small to notice they make a very noticeable difference in the behaviour of the pendulum. This is a form of chaos.***

* The same pendulum will also go through its entire repertoire of gyrations as it slows down, because although its speed is only changing gradually as it slows, at some points a small change in speed will flip it into a different pattern of oscillations.

chaos can be found throughout the natural world

KEY WORDS

SELF-ORGANIZED CRITICALITY: the way patterns appear in a system just before it tips over into chaos

CHAOS, COMPLEXITY AND LIFE

The most dramatic implications of chaos are only beginning to be investigated as we enter the new millennium. They link all of these ideas to the appearance of complexity in the Universe, and to the mystery of life and evolution. ***The key new discovery is that complex things exist 'on the edge of chaos', at the border between stability and chaotic behaviour.*** The neatest analogy to demonstrate this comes from PER BAK, a Danish physicist who describes complex systems as being in a state of 'SELF-ORGANIZED CRITICALITY'.

SELF-ORGANIZED CRITICALITY

* All you need to demonstrate self-organized criticality is a tray of sand. If sand is spread out all over the tray, it is in a stable - and boring - state. Nothing interesting happens to it. But if you drop grains of sand onto the middle of the tray, one by one or in a slow stream, they build up a pile.

hang on mum, I want to reach the critical point

SCALE-FREE PHENOMENA

Earthquakes, large and small, are all generated by the same process, just like the avalanches produced by adding grains to a pile of sand. There's nothing special about large earthquakes, except for their size. This is good news for physicists, because it means you only need one theory to describe all kinds of earthquakes, ***whatever their size.*** Another way of describing this kind of phenomenon is to say that it is 'SCALE FREE'.

DRAMATIC BUILD-UP

* At first, this is just as stable and almost as boring as the rest of the sand. But at a critical point, little landslides and avalanches start to occur in the growing pile of sand. You reach a stage where adding just one grain of sand triggers a lot of avalanches, rearranging the sand pile into an interesting and complicated structure. Then, as you keep adding sand to it, it will build up again, before collapsing in the same dramatic fashion.

chaos is scale free

But even when the sand pile is on the edge of stability, it doesn't always collapse completely – adding that extra grain of sand may merely trigger a small or medium-sized avalanche. It is impossible to predict in advance which kind of avalanche will happen next.

KEY WORDS

SCALE-FREE PHENOMENA: things that are the same except for their size are scale-free (a bonsai tree is a tree, even though it is small)

FREQUENCY AND SIZE

* *The collapsing sand pile is an example of a very common kind of pattern in nature – where similar events can occur on many different scales, but bigger events are less common than smaller events. Not just less common, but less common in a very precise way.*

* The frequency of an event of a certain size is inversely proportional to its size – or, turning this around, the size of an event is inversely proportional to its frequency (*f*). This is therefore known as **1/*f* noise**. 1/*f* noise turns up all over the place, in avalanches, earthquakes, traffic jams, and even in evolution.

an earthquake is the same thing as the collapsing sand pile, only bigger

Pure 1/*f* noise

A striking example of 1/*f* noise at work in the real world is the occurrence of earthquakes of different sizes. On the Richter scale, a magnitude 5 event is 10 times bigger than a magnitude 4 event, a magnitude 6 event is 10 times bigger still (and 100 times bigger than magnitude 4), and so on. It turns out that for every 1,000 earthquakes of magnitude 4 on the Richter scale, there are 100 with magnitude 5, 10 with magnitude 6, and so on. This is pure 1/*f* noise.

FEEDING OFF ENERGY

As we have just seen, a pile of sand can become a complicated and interesting structure on the edge of chaos. Yet each individual grain of sand automatically falls into place, in obedience to the laws of gravity and friction. A very simple set of physical rules has produced a relatively complicated pattern. There is, crucially, one other vital ingredient. We have been putting sand in from outside. ***This corresponds to a flow of energy through the system. What Bak and a few other physicists are saying is that simple laws feeding off a flow of energy can produce very complicated systems, without any other help at all.***

PUNCTUATED EQUILIBRIUM

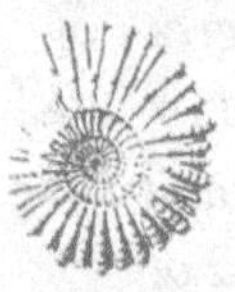

*** Some of the most exciting applications of these ideas involve life itself. Charles Darwin's theory of evolution by natural selection describes a gradual process, with tiny changes accumulating from one generation to the next. But some evolutionary biologists - notably the Americans** Niles Eldridge **and** Steven Jay Gould **- argue that the fossil record seems to show long periods with little change going on, alternating with short intervals in which a lot of evolutionary change takes place. This is known as** PUNCTUATED EQUILIBRIUM.

each individual grain of sand obeys the laws of gravity and friction

DARWIN'S VIEW OF EVOLUTION

* According to Darwin, evolution by natural selection occurs because of two things. First, genetic information is passed on slightly imperfectly from one generation to the next – so that the offspring of plants and animals resemble but are not exactly the same as their parents. Secondly, there is competition within any population for things like food or a chance to mate, so the individuals best 'fitted' to their environment are more successful and pass on more copies of their genes to later generations. The more copies of your genes you pass on, the more successful you are in evolutionary terms.
* Darwin's theory works beautifully to explain how organisms get progressively better at coping with the kind of environment they live in (which includes how they coexist or compete with other living things). ***But it is much harder to see how Darwin's theory on its own can account for the dramatic changes that occur from time to time, when many species disappear and then new species evolve (punctuated equilibrium).***

the evidence is in the fossil record

Darwin

Something extra?

Punctuated equilibrium does not mean that Darwin was wrong. But if what we are seeing in the fossil record is an accurate portrayal of what really happened, it means that something else needs to be added to Darwin's theory. ***That something else may be what we see in a growing sand pile, or in the historical record of earthquakes – the edge of chaos.***

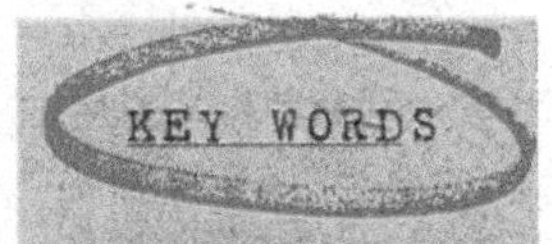

PUNCTUATED EQUILIBRIUM: when something stays the same for a long time, then undergoes a sudden change before resuming a state of stability

THE RED QUEEN EFFECT

*** Darwin's view of evolution implies that the whole web of life on Earth is constantly poised in an unstable state, like a growing sand pile. Everything in the ecosystem has to keep on evolving, to preserve its own niche, because everything else is evolving. This is known as the Red Queen effect, from the character in Lewis Carroll's *Through the Looking-Glass* who has to keep running as fast as she can in order to stay in the same place.**

the Red Queen has to move as fast as she can to stay in the same place

The peacock's tail

The splendid tail of the peacock may be a result of instability and the Red Queen effect. If – for whatever reason – peahens start to favour peacocks with large tails, then those peacocks will have more offspring than their less well-endowed brothers. Over many generations, this can produce the tails we see today.

EQUILIBRIUM RESTORED

well it's not eating me

* To see what biologists mean by the Red Queen effect, think of two different species living alongside one another. Suppose there are frogs that eat a certain kind of fly, which they catch by flicking out their tongues. If the frogs evolve a particularly sticky tongue, they will be adept at catching flies. The frogs will do well, and the flies badly, in the evolutionary stakes.

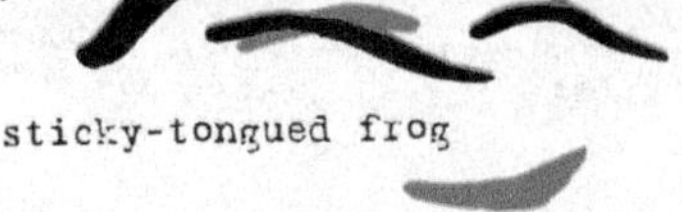

sticky-tongued frog

* But if the flies evolve a particularly slippery body surface, they will be able to escape from the sticky tongue more easily – and the original balance will be restored. Both species have evolved in the way Darwin envisaged, but neither species has benefited as a result. ***Overall, nothing has changed. There are still the same number of frogs, each of them eating the same number of flies.***

STABILITY, THEN INSTABILITY

* ***The Red Queen effect is in evidence all around us, all the time – involving many species interacting with each other and their physical environment simultaneously, not just pairs of species.*** Per Bak built a computer model of this kind of behaviour. As you might expect, competitive evolution encourages the spread of individuals that are best fitted to their environment (including the effect on them of other species). ***But once the happy situation is reached where all the species are well adapted, the system turns out to be unstable – in exactly the same way that happens with the growing pile of sand.***

the fossil record shows the patterns of extinction

PATTERNS OF EXTINCTION

When Bak made a small change to his model, removing just one species from his computer ecosystem, he found that sometimes this produced a small ripple effect as the other species adapted, sometimes it produced a big effect, and sometimes the ripples spread dramatically – with many species disappearing and others evolving to take their place. ***He had produced a model of punctuated equilibrium, a scale-free model obeying the 1/f law. What makes this discovery so special is that if you look at the fossil record and make a note of how often extinctions of a certain size have occurred, you find that this pattern follows a 1/f law, too.***

THE WEB OF LIFE

* According to Bak, life on Earth is an integrated web - the failure of one part of the system may have repercussions that affect the whole system. Maybe the flies that frogs feed on become so good at escaping that the frogs starve. So the fish that eat the frogs' eggs and tadpoles starve. Then the bears that feed on fish go hungry and start to eat rabbits (say) instead. And so on, in a series of widening ripples. It may be that the change is triggered by a change in the physical environment - a new rainfall pattern, or a burst of volcanic activity, or whatever. The important point is that extinctions of all sizes can be triggered by the *same* events.

I'm bound to catch something here if I wait long enough

Scale doesn't matter

As with earthquakes, there's no need for a different theory to explain extinctions on different scales. All you need is for something to trigger a small change in the web of life, which may or may not spread. Small meteorites hitting the Earth every so often could explain extinctions in the fossil record of life on Earth on *all* scales.

DINOSAURS, METEORITES AND CHAOS

* Some people think that the catastrophic extinction that occurred 65 million years ago was caused by a meteorite striking the Earth from space. Because this was a particularly big extinction, in which a huge number of species (including the dinosaurs) died, it is natural to think that it was caused by the impact of a particularly large meteorite. It's then logical to assume that smaller extinctions, in

which only a few species died, must have been caused by the impact of smaller meteorites. ***** ***But Bak's insight says that this need not be the case. Even a small meteorite crashing into the Earth is capable of triggering a massive extinction, in just the same way that adding a single grain of sand to the growing pile can trigger a catastrophic collapse. And it happens for the same reason – because both systems, the growing sand pile and life on Earth, are in a state of self-organized criticality, on the edge of chaos.***

THE EDGE OF CHAOS

Bak's computer models do not yet reflect the complexity of the real world. A lot more work needs to be done, building on his discovery. But it may mean that all the puzzling extinctions in the fossil record, the punctuation marks of evolution, are no more than the biological equivalent of adding another grain of sand to a pile in a critical state. When just one or two species become extinct, or develop a mutation that changes their place in the web of life significantly, the results may be small-scale, or medium-scale, or dramatic. ***Because life itself is poised on the edge of chaos.***

Jim Lovelock

Jim Lovelock

British chemist and inventor best known as the originator of the Gaia hypothesis, Lovelock also designed instruments used by NASA in the search for life on Mars and the highly sensitive detectors that revealed the way CFCs spread around the globe. Lovelock is that rare phenomenon, a truly independent scientist. Since 1964 he has lived off the income from his inventions and books while carrying out fundamental research in his own laboratory, at his home in the south of England.

THE GAIA HYPOTHESIS

* **Per Bak's view of life on Earth as a complex web of interacting species, intimately affected by changes in the environment, provides an intriguing underpinning for** Jim Lovelock's **idea of the Earth as a single living organism, called Gaia.**

A QUESTION OF ATMOSPHERE

* In the 1970s, when he was working for NASA on the experiments intended to search for life on Mars, it occurred to Lovelock that the Earth is in effect a single superorganism. Mars, he argued, must be a dead planet, because it is in a state of chemical equilibrium, with an atmosphere of inert carbon dioxide. ***The atmosphere of our planet, on the other hand, is in an unstable state, since it is rich in oxygen, a highly reactive gas.***

POISED ON THE EDGE

* The Earth is maintained in this state – on the edge of chaos – by life, which both produces the oxygen and depletes it. The process by which we breathe in oxygen and turn it into carbon dioxide is like a slow burning; and this prevents a major conflagration of the planet, which would occur if the oxygen level increased.

lung

we breathe in oxygen and turn it into carbon dioxide

the energy from the Sun is what sustains the self-organized criticality of life on Earth

* ***As with the growing sand pile, and all systems involving self-organized criticality on the edge of chaos, the whole Earth 'feeds' off a stream of energy from outside, in the form of heat and light from the Sun.***

THE END OF PHYSICS?

Some people claim the end is in sight for physics, because we have discovered all the simple rules by which the Universe operates. This is like saying that because you've learnt the rules for moving chess pieces, you are ready to take on a grandmaster. ***The truth is that, as the 21st century dawns, physicists are only just starting to understand the Universe.*** It has taken a little over 300 years, from the time of Isaac Newton to the present day, to work out the rules of the game. Now, we have to learn how to play it and to appreciate that complexity arises from simple beginnings in a Universe governed by simple laws.

plants change carbon dioxide into oxygen

Physics

Philosophy